Magnetic Domains

METHUEN'S MONOGRAPHS
ON PHYSICAL SUBJECTS

General Editors:

B. L. WORSNOP, B.Sc., Ph.D.
G. K. T. CONN, M.A., Ph.D.

Magnetic Domains

R. S. TEBBLE
Ph.D., D.Sc., F.Inst.P.
Professor of Physics
University of Salford

METHUEN & CO LTD
11 NEW FETTER LANE, LONDON EC4

First published 1969

Printed in Great Britain by
Butler & Tanner Ltd, Frome and London
SBN 416 10740 0

Distributed in the U.S.A. by
Barnes & Noble, Inc.

TO

E. C. STONER

Contents

Plates

(*between pp 20 and 21*)

Preface

A knowledge of the structure and properties of magnetic domains is essential to a real understanding of the magnetization process in ferromagnetism, and in this monograph I have endeavoured to present the elements of the subject.

This book is intended not only for students of the subject but also for physicists and engineers with an interest in magnetic materials. These materials were at one time used mainly in electrical machinery, transformers and permanent magnets, but with the production of new compounds and improved magnetic thin films there are now wider applications, for instance, to microwave devices and computer memory elements. With this has come an increasing awareness of the significance of the domain properties of these ferromagnetic materials.

Diagrams and micrographs have been used freely to illustrate the text; this is I feel appropriate as the pictorial approach, together with experimental observation, has played such a large part in the progress of domain studies.

R.S.T.

Acknowledgments

I should like to express my thanks to Dr Brian Johnson who read the manuscript, and to the authors and publishers, acknowledged in the text, who kindly provided me with the photographs used for the plates.

R.S.T.

1: Introduction

Many substances can be considered as being magnetic in the sense that there may be associated with each individual atom or molecule a magnetic moment. This moment (μ) may arise either from the orbital motion of the electron or may be associated with the electron spin. We are concerned here, however, with ferromagnetic materials where in the simplest case the spin moments are aligned parallel to each other so that there is in the specimen a resultant magnetic moment, even in the absence of an external applied magnetic field (Fig. 1.3a). The parallel alignment of the spins is due to the quantum mechanical exchange interaction between electron spins on adjacent atoms tending to align the spins parallel or antiparallel to each other. The exact mechanism of the interaction is not relevant here but the effect is equivalent to that of an internal magnetic field acting on the individual atomic moment, of the order of 10^7 oersted.

This interaction field can be represented as a pseudo-magnetic field, the Weiss field H_W acting so as to align the spin moments and proportional to the spontaneous magnetization I_s, i.e. $H_W = N_W I_s$. This alignment is complete at absolute zero when, with n atoms per cm^3, I_s takes its maximum value $I_0 = n\mu$. With increasing temperature the alignment is disturbed by thermal agitation so that the spontaneous magnetization I_s decreases with increasing temperature as indicated in Fig. 1.1 until at θ_C the Curie temperature, the system is disordered completely (Fig. 1.3b). The relation between I_s and T is given for a system with spin $S = \frac{1}{2}$ by

$$I_s/I_0 = \tanh(\mu H_W/kT) \tag{1.1}$$

where k is Boltzmann's constant.

Thus both I_s and H_W decrease with temperature.

Despite the presence of these large interaction effects tending to align the spin moments, it is found that in most ferromagnetic materials large changes in magnetization can be brought about by the application of relatively small applied fields, often of the order of only a few oersteds, and the specimen can be demagnetized in

fields of similar magnitude. This behaviour may be summarized in the hysteresis curve shown in Fig. 1.2. On applying a field to the demagnetized specimen, the magnetization changes at first slowly and reversibly and then more sharply and irreversibly at higher fields until it reaches a value approaching saturation. On reducing the field to zero, the magnetization falls to a remanence value I_r; a reverse

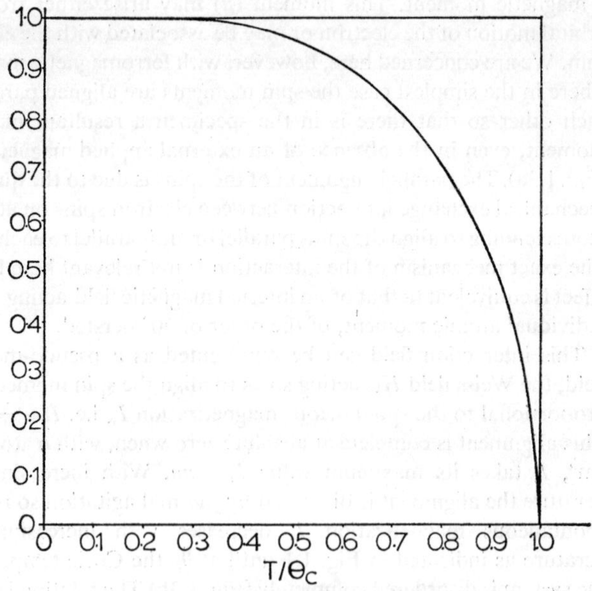

FIG. 1.1. Curve of reduced magnetization I_s/I_0 against temperature T; θ_C is the Curie temperature.

field (the coercive field or coercivity H_c) is required to reduce the magnetization to zero; the cycle continues round the hysteresis loop as shown. In the steep part of the curve the magnetization changes discontinuously in a series of small steps as shown inset; the changes in magnetization are reversible over parts (a) and discontinuous and irreversible over parts (b). The discontinuous nature of these changes is demonstrated in the Barkhausen effect. The apparatus consists of the standard arrangement for measuring hysteresis loops with a magnetizing coil and a search coil surrounding a suitable specimen, such as an iron wire; the usual galvanometer is replaced by an amplifier followed by an oscilloscope or loudspeaker. On changing the magnetic field slowly over the steep part of the magnetization curve, the small voltage pulses induced in the search

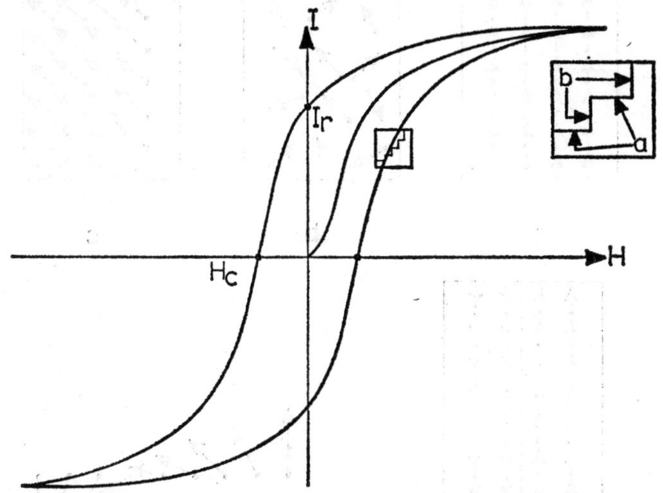

FIG. 1.2 (a–b) A typical hysteresis curve with inset a much enlarged section of the curve showing: (a) reversible, and (b) irreversible changes in magnetization. I_r is the remanence and H_c the coercivity.

coil by the discontinuous changes in magnetization can be observed individually on the oscilloscope or detected as clicks on the loudspeaker. The effect is observed no matter how slowly the applied field is changed. The discontinuities are at a maximum in number and size in the steep part of the hysteresis curve, near the coercivity, and decrease towards saturation in the high field region where the changes in magnetization are reversible.

It is with hysteresis and the magnetization process in low fields that domain structure is associated. In order to introduce this, let us

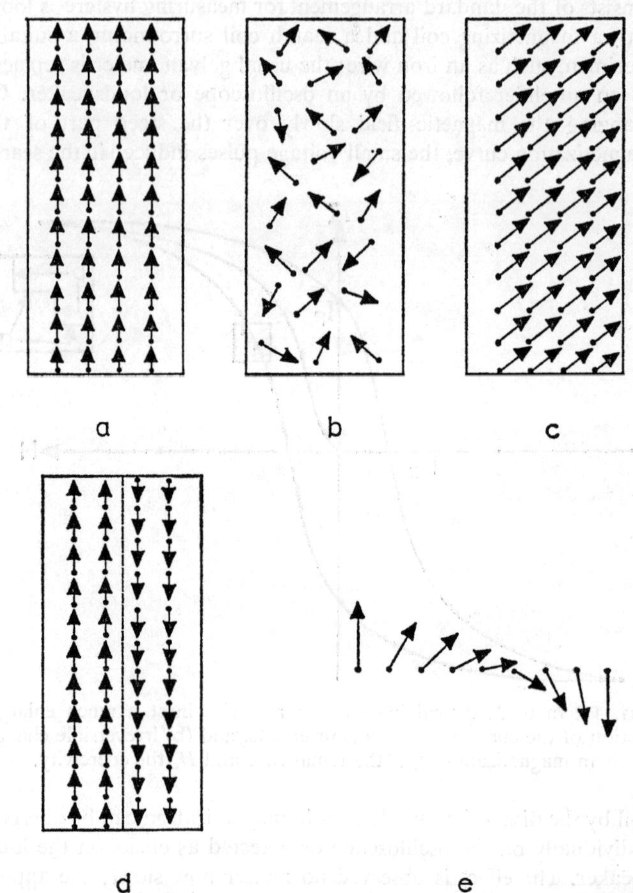

FIG. 1.3 (a–e) Representation of atomic moments in a single crystal: (a) spontaneously magnetized at 0°K with spins parallel; (b) with spins disordered above the Curie temperature, (c) with coherent rotation of spins, (d) with spins antiparallel in two 180° domains, (e) arrangement of spins in a 110° domain wall.

consider first of all the difficulties involved in reversing the magnetization in a ferromagnetic specimen such as the single crystal represented in Fig. 1.3a. In order to reverse the magnetization in such a specimen the following methods are amongst those possible: (i) the orientation of the spins may be disordered and re-aligned in the opposite direction, or (ii) the spins may be rotated in unison to point in the opposite direction.

The first method (i) requires that each individual spin be subject to a reverse magnetic field of the order of 10^7 oersteds so as to produce the required rotation, or that the specimen be heated to the disordering temperature known as the Curie temperature and cooled with a field applied in the reverse direction. In the second procedure (ii) that of rotating the spins in unison, we come up against the effect of demagnetizing fields that arise at the surface of the specimen, i.e. at the interface between magnetic and non-magnetic media. The specimen is shown as magnetized in the long direction where relatively low demagnetizing fields prevail; this is the state of minimum energy (Fig. 1.3a). However, on rotating the magnetization (Fig. 1.3c), the demagnetizing effect would increase to an extent dependent on the shape of the specimen and might require applied fields of the order of several hundred oersteds to overcome.

The process of reversing the magnetization is made easier if the specimen is divided magnetically into domains or regions, each magnetized to saturation (Fig. 1.3d). A simple example is shown in Fig. 1.4a in which the overall magnetization is zero. Upon applying a magnetic field in say the $+z$ direction the domain magnetized parallel to the $+z$ direction, i.e. parallel to the field, grows at the expense of the other by movement of the boundary between the domains. In this example the movement is to the right (cf. Plate N).

The magnetic field required to bring about the boundary movement is relatively small and in the example shown is that necessary to balance the increase in the demagnetizing field at the ends of the specimen. At sufficiently high fields the boundary reaches the edge of the specimen, but even in high fields some rudimentary domain structure is retained so as to form a nucleus for the re-formation of the original structure. The whole process can be described in terms of a curve of internal energy against displacement x; on applying a field the boundary moves up the energy curve to a position of equilibrium and returns as the field is reduced (Fig. 5.1b).

This is a reversible process whereas in fact the boundary motion is

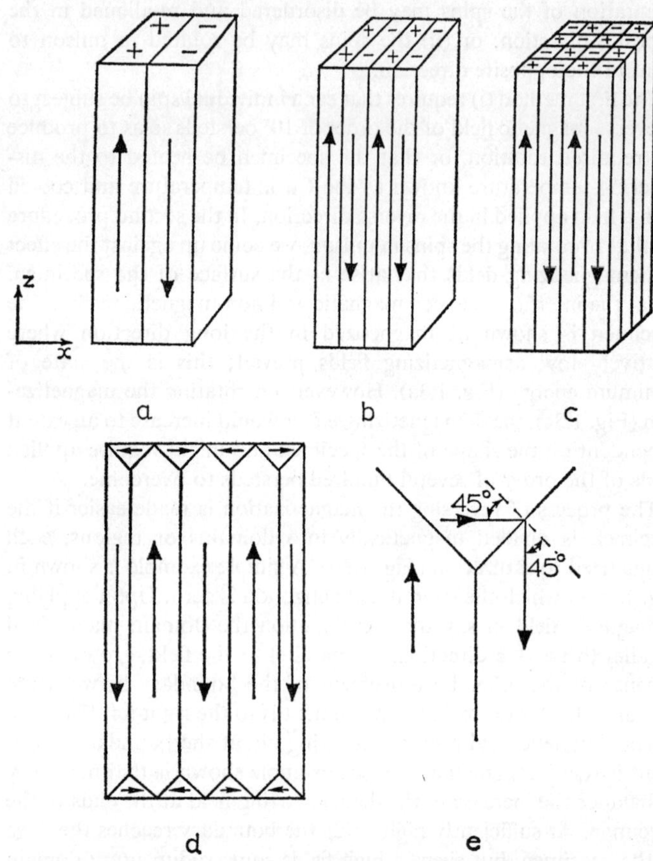

FIG. 1.4 (a–d) Examples of domain structures. (e) Directions of magnetization vectors at the 90° boundary of a closure domain. (f) Formation of reverse spikes of magnetization. (g) Honeycomb pattern, an alternative to the checkerboard of Fig. 1.4(c).

impeded by imperfections in the crystal structure so that the magnetization process is irreversible. The irreversible effect is represented on the energy (E, x) diagram showing a curve with a number of maxima and minima. The condition for equilibrium is the same as before; when a magnetic field is applied the boundary moves to a position of equilibrium but when the boundary reaches a point where $\mathrm{d}E/\mathrm{d}x$ is a maximum the system becomes unstable and the boundary moves discontinuously to a new equilibrium position (Fig. 5.1d). The boundary motion thus consists of a series of steps corresponding to the Barkhausen discontinuities in the magnetization curve. It is shown later (Chapter 5) that on reducing the field the return path is not the same and the field must in general be reversed to bring the boundary along another path; the result is that a miniature hysteresis loop is traversed. The hysteresis curve for the specimen as a whole is the sum of these individual processes.

There are, of course, more complex domain structures than the simple configuration so far considered. The energy can be reduced by further subdivision of the structure as shown in Fig. 1.4b. This brings about a reduction in the magnetostatic energy associated with the demagnetizing fields by introducing an alternation in the sign of the 'free pole' arising at the ends. There is a limit to this

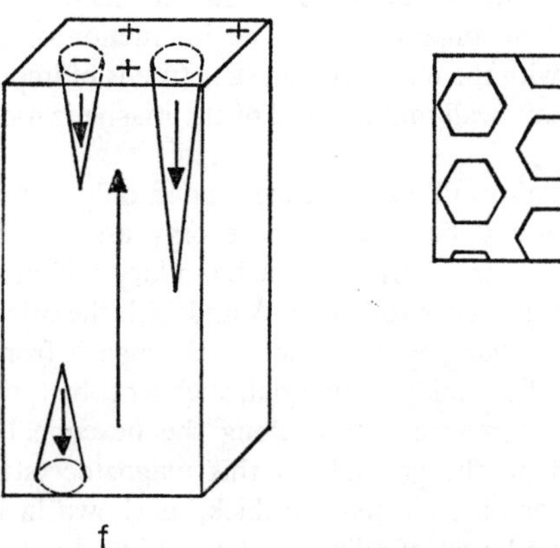

FIG. 1.4 *continued*

continued subdivision because of the additional energy introduced by domain wall formation, and the final dimensions of the individual domain represent a balance between the two effects.

Most ferromagnetic materials possess a preferred direction of magnetization parallel to one or other of the major crystallographic axis. In hexagonal cobalt this is parallel to the *c* axis and much of what has been said above applies to a hexagonal single crystal with the *c* axis parallel to the *z* coordinate in Fig. 1.4. In iron, for comparison, there are easy directions of magnetization which lie parallel to the cube axes, and a domain structure such as that in Fig. 1.4d may be formed in the demagnetized state. Note (i) that the magnetization lies parallel to the ⟨100⟩ crystallographic directions both in the main domains and in the 'closure' structure at each end and (ii) that across any domain boundary the normal component of the magnetization is continuous so as to avoid the production of localized magnetic fields. Deviation from either of these conditions gives rise to relatively large contributions to the energy of the system, and the formation of any domain configuration is to a very large extent controlled by the need to minimize the energy due to these two factors. More involved domain configurations arise at inclusions, and at imperfections in the crystal structure, and these usually form in such a way as to reduce the effect of the localized magnetic fields or 'free pole' arising at these imperfections. The interaction of these subsidiary domains with the main domain structure is of importance in the study of domain wall motion, and of the magnetization process generally.

So far the domain wall has been mentioned only as a boundary between domains and because of the energy associated with its formation; it is better described as a boundary region extending over distances of the order 10^2 to 10^3 Å and with the orientation of the spin moments changing continuously through it from one domain to the next. In a uniaxial material, such as cobalt, with a preferred direction of magnetization along the hexagonal axis, the boundary would on the grounds of this magnetocrystalline anisotropy be only one atomic spacing thick, as shown in Fig. 1.3d; however, with two layers of adjacent atoms aligned antiparallel to each other there would be a large contribution from the exchange interaction forces tending to align the spins parallel. The spin vectors are therefore orientated as shown in Fig. 1.3e so that the wall structure represents a balance between the magnetocrystalline anisotropy and the exchange interaction; there is then not only an energy but an effective thickness to be associated with the domain boundary.

This is of some importance not only because of the contribution of domain-wall energy to the total energy of a specific domain configuration but also in the detailed magnetic structure arising at very small inclusions and imperfections in the crystal structure. In the special case of thin films, where the thickness of these is comparable with the domain-wall thickness, the boundary structure may be modified so as to alter quite radically the geometry of the boundary configuration. For instance, in very thin films, cross-tie walls and 'Néel' walls may be formed in place of the 'Bloch'-type boundary observed in larger specimens.

There are a number of methods of observing magnetic domains but only three are of importance; these are the Bitter or powder pattern technique, transmission electron microscopy and magneto-optic methods. In the Bitter method (and the Craik modification) the surface of a carefully polished crystal is covered with an aqueous colloidal suspension of magnetite Fe_3O_4; the magnetic oxide particles collect at domain boundaries, where the field gradient is a maximum, forming a pattern outlining the domain structure (see Plate A). With the electron microscope, use is made of the Lorentz force acting to displace the electron beam; details of the imaging process are not straightforward and these are discussed later. The method can be applied only to thin films or foils and is therefore of limited application, but it has a much higher resolution than any other method (Plate P). Domains may also be observed by their effect upon the reflection of plane polarized light incident on the crystal surface; this is, to a sufficient approximation, a rotation of the plane of polarization proportional to the intensity of magnetization near the surface. An example of this (the Kerr effect) is shown in Plate C; the transmission (Faraday effect) has also been used (Plate D).

The development of experimental methods of observation is of even greater importance in the study of domains than in many other branches of physics as it is not possible to derive a given domain structure from a general solution of the appropriate equations. The general method of tackling the problem is to find from experimental observation what specific domain structures are formed under certain conditions and to derive the appropriate energy expressions; the conditions under which their formation takes place and their behaviour in a magnetic field can then be 'theoretically' analysed in terms of a possible 'model'. There are in fact a number of instances in which the formation of a particular configuration has been postulated before being observed experimentally. It should be possible to derive the magnetization distribution corresponding to known domain configurations from first principles although this has not yet been achieved. The way to the development of the necessary techniques (Micromagnetics) has been pointed out by W. F. Brown (1962) in his book *Magnetostatic Principles in Ferromagnetism.*

One aspect of the subject has not been mentioned; this is the problem of how domains are formed initially. Although a particular domain configuration may be 'justified' on the grounds that the energy associated with it is lower than any other conceivable structure, the process of its formation or nucleation may involve a relatively high energy. The rotation of electron spins necessary to produce a domain wall requires large localized magnetic fields, much higher than would be expected even in the regions of flaws or sharp corners in the specimen. This problem has not yet been satisfactorily solved and the subject is discussed in a later chapter.

1.1 Ferromagnetic materials

The important ferromagnetic elements are iron, cobalt and nickel; some of the rare-earth metals such as gadolinium are also ferromagnetic at low temperatures but these do not concern us here. The significant properties of iron, cobalt and nickel are summarized in Table 1.1; iron and cobalt have a high saturation magnetization and a high Curie point with much lower values for nickel. Cobalt has a hexagonal crystal structure and possesses a strongly preferred direction of magnetization parallel to the hexagonal axis, i.e. it has high magnetocrystalline anisotropy; at temperatures above ~420°C

Table 1.1 Room temperature properties of ferromagnetic metals and oxides

	I_s (e.m.u. gm^{-1})	$K \times 10^{-3}$ (erg cm^{-3})	H_K (oe)	$\lambda_s \times 10^6$	θ_C (°C)	$A \times 10^6$ (erg cm^{-1})
Iron	1710	480	560	−7	770	2·0
Cobalt	1430	5300	7400	−50	1120	1·3
Nickel	485	−45	185	−34	360	0·75
Magnetite Fe_3O_4	480	−110	460	+40	585	
Manganese ferrite $MnFe_2O_4$	400	−40	200	−5	300	
Cobalt ferrite $CoFe_2O_4$	425	2000	4700	−110	520	
Nickel ferrite $NiFe_2O_4$	270	−62	230	−26	585	
Barium ferrite $BaFe_{12}O_{19}$	380	3300	17,000		450	0·61

it becomes face-centred cubic like nickel. Iron and nickel have cubic crystal structures, iron (bcc) with a preferred direction along the cube edge and nickel (fcc) along a cube diagonal. Nickel has a low anisotropy but is highly magnetostrictive, and its magnetic properties are very susceptible to strain.

Of the alloys, silicon iron (2 to 4% silicon) is of importance industrially in electrical machinery and transformers because of its low hysteresis and high permeability. The nickel-iron alloys in the range 45 to 80% nickel (Permalloy, Mumetal) are widely used in transformers in the communications industry because of their high permeability. For, although they have a relatively low magnetization, they have low magnetostriction and low magnetocrystalline anisotropy which is practically zero near 80% nickel. These alloys are increasingly used in magnetic thin films for computer memories. The permanent magnet alloys are discussed in section 6.3.

The ferrites.

These are the mixed oxides, with the general formula $M^{2+}O.Fe_2O_3$ where M is a metallic ion. The best-known example is magnetite Fe_3O_4, the lodestone of the ancients; in terms of the above formula it is $FeO.Fe_2O_3$. The ferrites are of importance technologically because as non-metals they have a high electrical resistivity so that eddy-current losses are low and they can be used at high frequency; because of its low resistivity the exception is magnetite. They are used not only in telecommunications work but also in memory stores in computers.

The ferrite crystal structure is that of the spinels; this consists of an fcc lattice of oxygen ions with metallic ions lying either at the centre of the cube edge (octahedral B site) or on a cube diagonal on a tetrahedral A site. The structure is more or less the inverse spinel; one half of the iron ions lie on B sites and the other half, together with the M ions, lie on A sites. The spins on A-site ions lie antiparallel to the B-site ions and the imbalance in the spins leads to a net magnetic moment. This phenomenon is known as ferrimagnetism, but so far as domain structure is concerned there is no significant difference between ferrimagnetic and ferromagnetic materials. The hexagonal ferrites are described in section 6.3.

Other oxide materials include the rare-earth garnets which have the formula $Fe_5M_3O_{12}$, i.e. $5Fe_2O_3.3M_2O_3$, where M is yttrium or a rare-earth ion. The garnets have higher resistivities than the spinel oxides and can therefore be used at higher frequencies. Some are transparent to light and some use has been made of these magneto-optical properties for 'switching' light beams. Their Curie temperatures are about 550°C and some have compensation points just above room temperature; this last is the temperature at which the moments on the different lattices cancel each other, although the moment on each lattice does not fall to zero until the Curie temperature is reached.

2: Methods of Domain Observation

A number of methods of observing domains have been devised (see Craik and Tebble, 1965) but only three are of real significance in the study of domain structure; the Bitter method (with the Craik modification) which historically has been by far the most important, the magneto-optic methods and Lorentz electron microscopy.

2.1 The Bitter method

The Bitter method is a refinement of the well-known demonstration using iron filings to map out the magnetic field in the vicinity of a bar magnet; in this technique a fine colloidal suspension of ferromagnetic Fe_3O_4 (magnetite) is spread over the surface of the specimen. A precipitate of magnetite (Fe_3O_4) is prepared by adding 250 c.c. of 10% sodium hydroxide solution (w/w) drop by drop to 300 c.c. of a solution containing a mixture of 2 gm $FeCl_2 4H_2O$ and 5·4 gm $FeCl_3 6H_2O$; this is heated and vigorously stirred. The precipitate is filtered and washed repeatedly in distilled water. The filtrate is then washed with 0·01N HCl then added to 0·5% soap solution (sodium oleate) and boiled until a permanent suspension is produced.

A small drop of the liquid is placed on the surface of the specimen. Under the action of the gradient in the stray magnetic fields, the colloid particles are attracted to the regions of maximum magnetic field which coincide with the lines along which the domain boundaries intersect the surface of the specimen (Fig. 2.1). Thus a pattern is formed on the surface of the specimen outlining the domain boundaries and of course any other region of magnetic inhomogeneity which may arise around inclusions and imperfections. A typical example is shown in Plate A.

In many cases, in addition to the main domain configuration, the Bitter pattern appears to have a fine structure consisting of striations lying at right angles to the direction of magnetization. This is caused by the aggregation of colloid particles at cracks in the surface giving rise to stray magnetic fields which are a maximum where the cracks lie at right angles to the magnetization (Fig. 2.1 and Plate I). These striations thus provide a means of determining the direction of magnetization in the domain pattern; (with improved methods of polishing, the surface of the specimen may be so free from blemishes that it may be necessary to scratch the surface to produce the effect). In practice this direction can often be determined from a consideration of the domain geometry, bearing in mind the requirement that

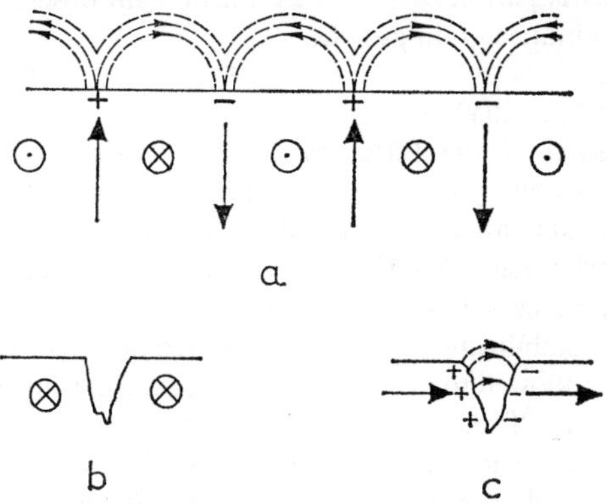

FIG. 2.1 (a–c) Bitter pattern formation: (a) the fields arising at the intersection of domain walls at the surface of a specimen, (b) and (c) the stray fields arising at cracks in the surface parallel and at right angles to the magnetization.

the free-pole energy should be a minimum. Bitter patterns are sometimes seen to consist of alternate bands rather than lines of colloid. These are formed where there is a component of the domain magnetization, I_n normal to the surface of the specimen. With the application of a magnetic field H_n normal to the surface the colloid particles are polarized and collect at the surface of those domains where H_n and I_n are parallel, usually at alternate domains, thus forming bands of colloid rather than lines. This effect can be observed in Plate I.

The preparation of the specimen calls for considerable care as the effect of strains can completely mask the magnetic features, a point which was not realized in the very early work. The surface, after suitable mechanical polishing, is electrolytically polished so as to remove the strained surface layers; it is sometimes necessary to protect the surface from the action of the colloid by covering it with a very thin layer of varnish or collodion. It is usually necessary for the specimen to be in the form of a single crystal, although patterns may be observed on carefully selected polycrystalline material, especially if it contains large crystallite grains. Silicon steel transformer sheet is particularly suitable for domain observation if it is annealed to induce grain growth.

2.2 The Craik method

The resolution of the Bitter method is restricted by the usual optical limits and by particle size to about 10^4 Å, so that the more detailed domains structure cannot be observed. Craik has developed a replica technique using a suspension of very fine colloid particles which may be as small as 200 Å in diameter. The colloid which contains a stabilizing agent, Cellacol (sodium carboxy methyl cellulose), is allowed to dry on the surface of the specimen so that the Bitter pattern is 'fixed' as the particles set in a solid film of Cellacol. The film is then stripped from the surface of the specimen and forms a replica suitable for observation by transmission electron microscopy. These replicas reproduce only static patterns so that continuous changes in domain structure cannot be observed, but this is more than compensated for by the improvement in resolution; see Plate B where the individual colloid particles can be observed.

2.3 Magneto-optic methods.

These methods make use of the effect of the domain magnetization on the polarization of plane polarized light observed by reflection from the surface of the specimen (the Kerr effect), or by transmission through the specimen (the Faraday effect). Both of these methods have the advantage that the changes in the domain pattern can be followed and that the specimen can be heated or subject to strain during observation. However, due to the generally small order of the optical rotation, there are experimental difficulties which have limited the wider adoption of the techniques.

(i) The Kerr effect

Plane polarized light on reflection from a magnetized surface is elliptically polarized. However, reflection from any metallic surface will, in general, introduce ellipticity except where the plane of polarization is either parallel or perpendicular to the plane of incidence; the reflected beam is then also plane polarized. It is necessary, therefore, in the application of the Kerr effect to ensure that the incident beam is oriented in one of these ways in order that the effect of the domain magnetization may be isolated from other reflection processes. In practice the degree of ellipticity due to the magnetization is small and can be looked upon as a rotation of the plane of polarization. The rotation produced is, at most, a few minutes of arc. The experimental techniques are consequently anything but straightforward and involve rather more than a simple system of polarizer and analyser, with at least the use of monochromatic light and the application of blooming techniques to the surface of the specimen. Examples of the results which can be obtained are shown in Plate C.

(ii) The Faraday effect

In this method the rotation of the plane of polarization of light transmitted through the specimen is due to the component of the magnetization parallel to the direction of the incident light. The specimens must be thin enough to allow transmission of the light and at the same time thick enough to produce a detectable rotation of the plane of polarization, which may be of the order of 10^2 degrees per mm. Metallic specimens are usually about 1000 Å thick, but with some of the oxide materials thicknesses of the order of 1 mm can be used. An example is shown in Plate D. Note that in both the Faraday and Kerr methods, it is the domain, rather than the domain wall, which is imaged.

2.4 Lorentz electron microscopy

The high resolution of the transmission electron microscope of the order of a few Ångstroms has led to a number of attempts to make use of the Lorentz force experienced by electrons traversing a magnetized specimen. Referring to Fig. 2.2, an electron beam passing normally through a thin ferromagnetic film magnetized parallel to the plane of the film is subject to a force given by

$$F_x = m\ddot{x} = \frac{-4\pi e}{c}(\mathbf{v} \times \mathbf{I}) = \frac{4\pi e v I_y}{c} \tag{2.1}$$

giving a deflection

$$\psi = \frac{\mathrm{d}x}{\mathrm{d}z} = \frac{\mathrm{d}x}{\mathrm{d}t}\cdot\frac{\mathrm{d}t}{\mathrm{d}z} = \frac{4\pi e t I_y}{mvc} \tag{2.2}$$

in an obvious notation. Thus, where the magnetic structure of the film consists of two antiparallel domains, as shown, the beam is

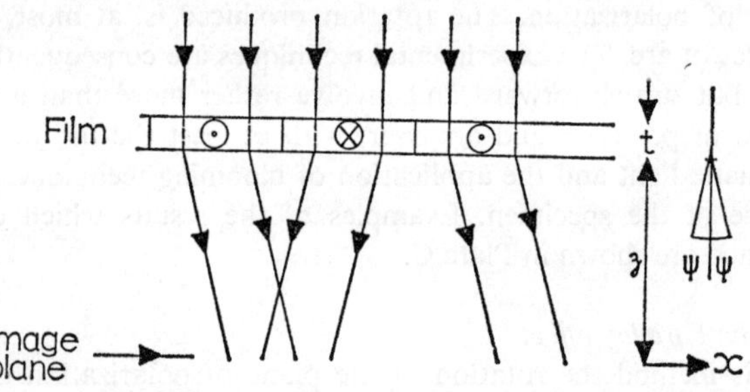

FIG. 2.2 Deviation φ of an electron beam passing through a ferromagnetic thin film thickness t. The domain structure consists of 180° domains (cf. Plate E) and the image is formed a distance z below the film.

divided and subsequently diverges or converges according to the sense of the magnetization.

The specimen obviously must be thin enough to be transparent to the electron beam, and for 100 keV electrons the maximum thickness for the ferromagnetic metals is about 2000 Å. Taking t = 2000 Å, I_y = 1700 e.m.u. cm^{-3}, and for 100 keV electrons the velocity $v = 1{\cdot}65 \times 10^{10}$ cm^{-1}; this gives ψ as approximately 4×10^{-4} radian.

Although the electron microscope may be considered in terms analogous to those of the optical microscope (Fig. 2.3), there are important differences, some of which are of particular significance in the present applications. One of these is the importance of scattering in producing contrast in image formation. Those areas of an object which scatter light so that it passes outside the aperture of the microscope correspond to areas in the image that are dark, and contrast in the image is produced by variations in the scattering. Clearly the effectiveness of the scattering process depends on the aperture of the microscope objective. The smaller the aperture the more effective would be the scattering in producing contrast.

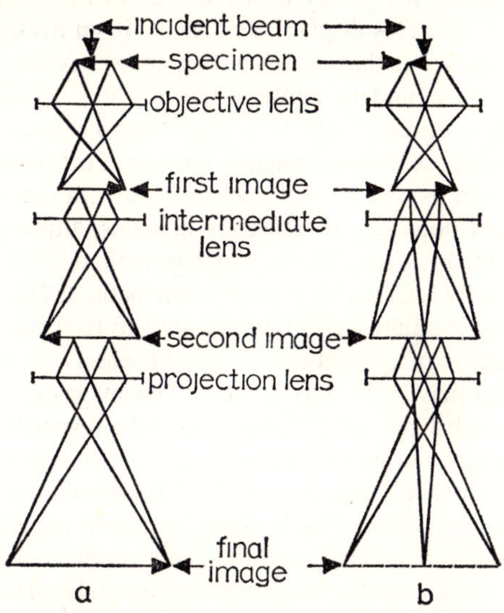

FIG. 2.3 (a–b) Diagram of electron microscope used: (a) for normal image observation, and (b) for diffraction.

The optical microscope has a relatively wide angular aperture, of the order of unity, and is relatively insensitive to scattering; image contrast with transmitted light is produced mainly by absorption or, using special methods, by phase contrast. Only in dark-field viewing, as for example when the centre of the aperture is filled with a circular stop, is scattering of importance.

In comparison the aperture of the electron microscope is very much smaller, typically 10^{-2} radian, and scattering is much more effective in producing contrast; it is in fact the main means by which contrast in image formation is produced in electron microscopy.

It is thus obvious that the order of magnitude of the scattering produced in magnetic domains, 10^{-4} radian, is much too small to produce any difference in intensity with an electron microscope objective of aperture 10^{-2}. If, however, the objective is focussed on a plane a distance z above or below the specimen the effect is to displace the domain images by a distance $z\psi$ in the x direction, as shown in Fig. 2.2. This displacement results in an overlap or divergence of the images so that there is a sharp increase or decrease in the intensity of the boundary between the domains, as illustrated in Plate E as light or dark lines. It is important to emphasize that the finite width of these lines is primarily a result of the displacement of the domain images and is not to be confused with the width of the actual domain boundary between the domains, although this may modify the image. This, the defocussing method of observing domains, is usually referred to as the Fresnel method. The defocussing distance z may be quite large, typically 10^{-2} to 10^{-1} cm so that with $\psi = 10^{-4}$ and a magnification of 10^4 the observed width of the boundary line may be 10^{-2} to 10^{-1} cm. The image of the boundary line is alternately light and dark because of the reversal in direction of the magnetization in successive domains; a change in the focussing plane from one below to one above the plane of the specimen results in a reversal in the intensity distribution. Although the objective is viewed 'out of focus' the depth of focus of the electron microscope is sufficient for much of the physical detail of the surface of the specimen to be observed (Plate P).

An alternative means of observing domains is the 'in focus' (Foucault) method. In this, use is made of the fact that in the diffraction pattern observed in the electron microscope the effect of the splitting of the electron beam by the antiparallel domain structure (Plate F) is to split a single diffraction spot into two. Each corresponds to the part of a beam passing through domains magnetized in one of the two directions. By displacing the objective aperture one of the components of the beam can be cut off and this is shown by the disappearance of one of the two diffraction spots. On refocussing the microscope for in-focus observation those

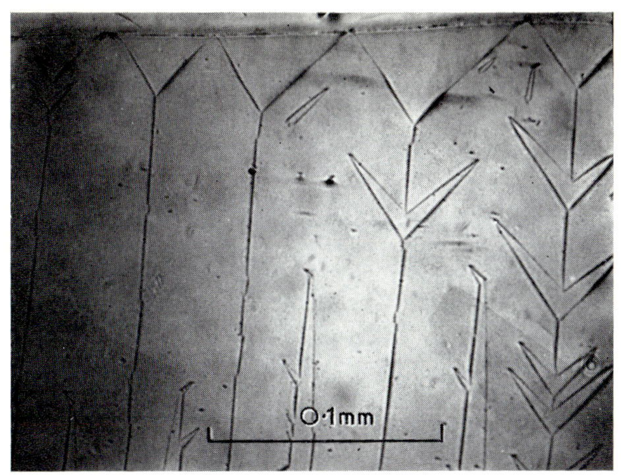

A. Bitter pattern of 180° domains and 90° closure domains in silicon iron (cf. Fig. 1.4). The easy < 100 > directions are ↑ →. Fir tree structures are formed as the surface is slightly away from (100) (*Dr W. Andra, private communication*) *pages* 9, 14, 39

B. Bitter pattern – Craik technique. Domain pattern on basal plane of barium ferrite; magnetization normal to surface. (*Dr D. J. Craik, private communication*) *pages* 16, 38

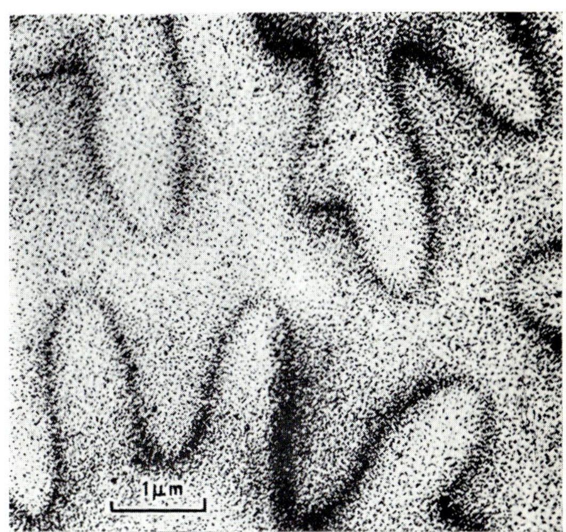

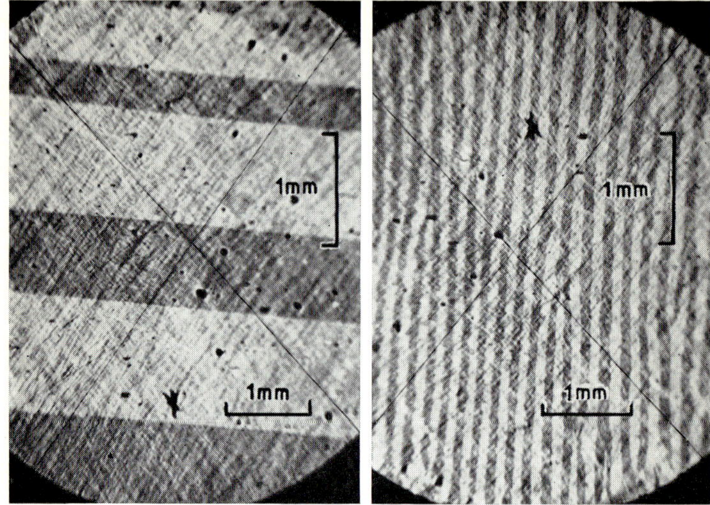

C. Kerr effect. (*i*) 180° domain structure on the (110) surface of silicon iron; there is a single [001] easy axis ↔ in the surface with the domain magnetization ↔. (*ii*) As in (*i*) but with a compressive stress ↔ producing an easy axis at right angles, i.e. ↕, so that the 180° domains lie ⇅ with the domain magnetization ⇅ cf. section 3.8. (*Dr J. Robey, private communication*) *pages* 9, 17, 33

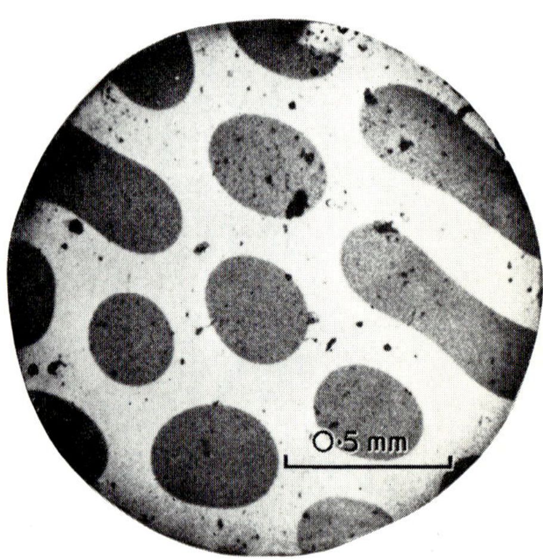

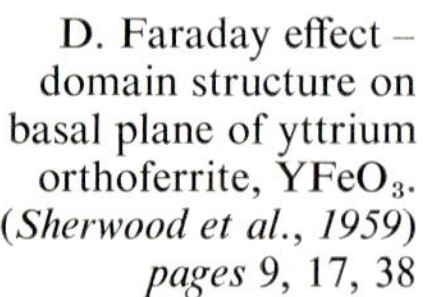
D. Faraday effect – domain structure on basal plane of yttrium orthoferrite, $YFeO_3$. (*Sherwood et al., 1959*) *pages* 9, 17, 38

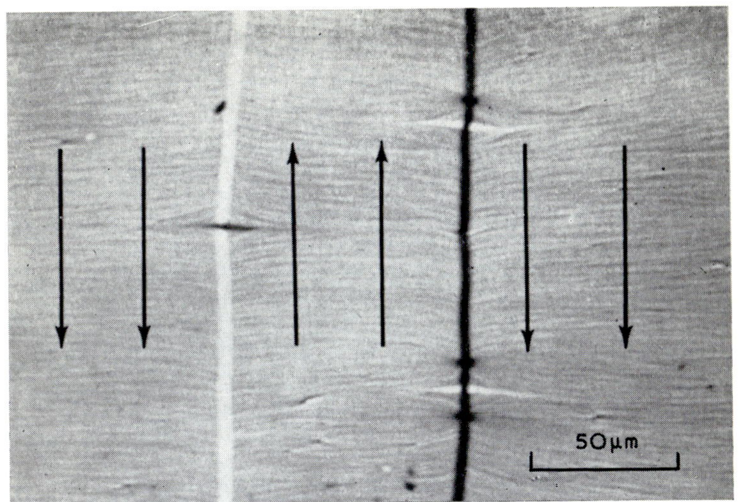

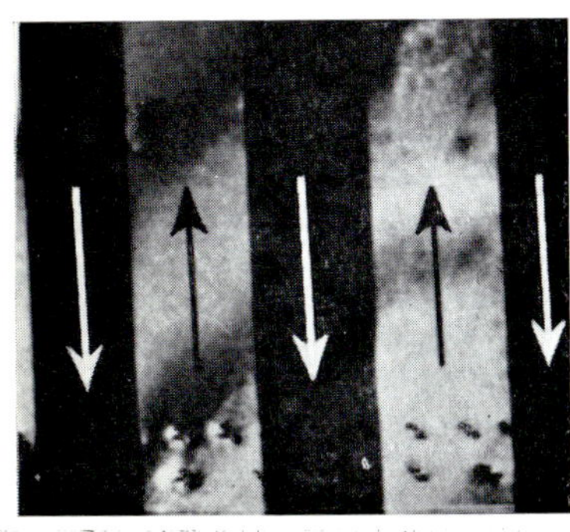

E. 180° domains in a thin film of Permalloy imaged by Lorentz microscopy – out of focus. (*Cohen, 1965a*) *pages* 20, 81

F. 180° domain in a thin film of iron imaged by Lorentz electron microscopy – in focus. (*Foucault method*) (× 2000) *page* 20

G. Interference fringes produced by electron beam passing through two 180° domains – 'Fresnel biprism effect'. (*Dr D. C. Hothersall, private communication*) *page* 21

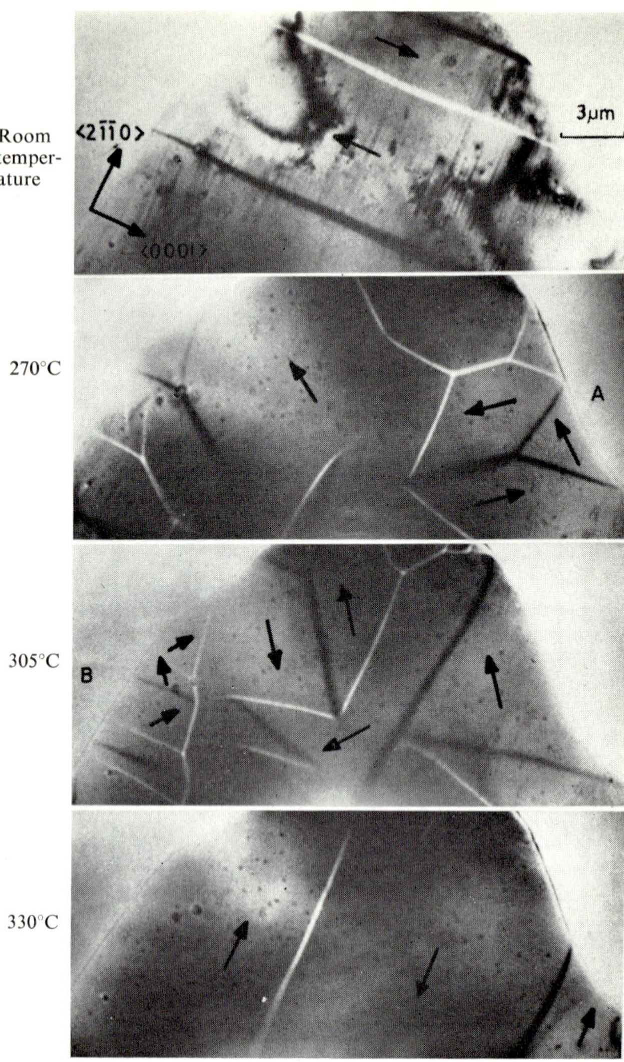

H. Change in domain structure in cobalt with temperature. (*Grundy and Tebble, 1968*) *page* 41

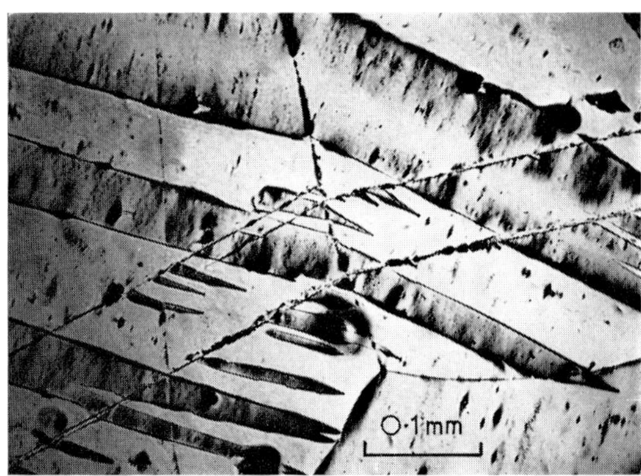

I. Bitter pattern as observed on the surface of silicon iron: note the change in direction of the domain walls and of the domain magnetization across the grain boundary (the colloid collects at cracks lying at right angles to the magnetization). Applied field H = 100 oersteds normal to the surface. (*Dr W. Andra, private communication*) *pages* 15, 39, 43

J. Cross-tie walls and Bloch lines in a thin film of Permalloy. (*Feldtkeller, 1961a*) *pages* 79, 81

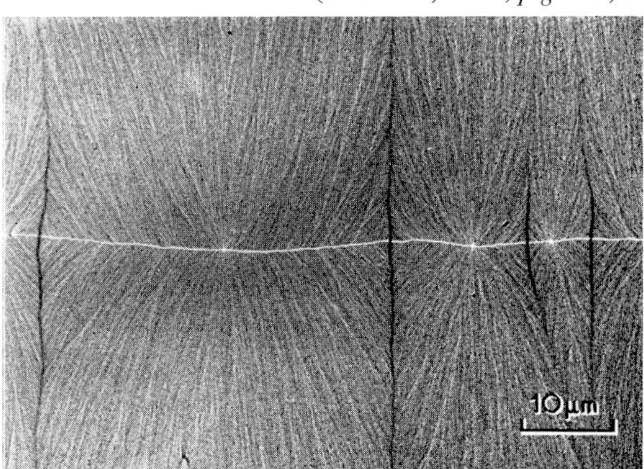

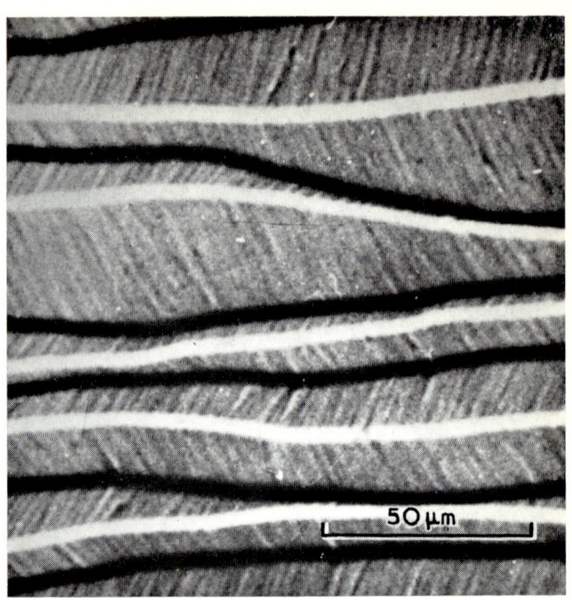

K. Domain structure in a thin film of Permalloy magnetized normal to the easy axis. (*Cohen 1965a*) *pages* 81, 84

L. Domain structure in an 'inverted' thin film of Permalloy. (*Feldtkeller, 1961b*) *pages* 81, 82

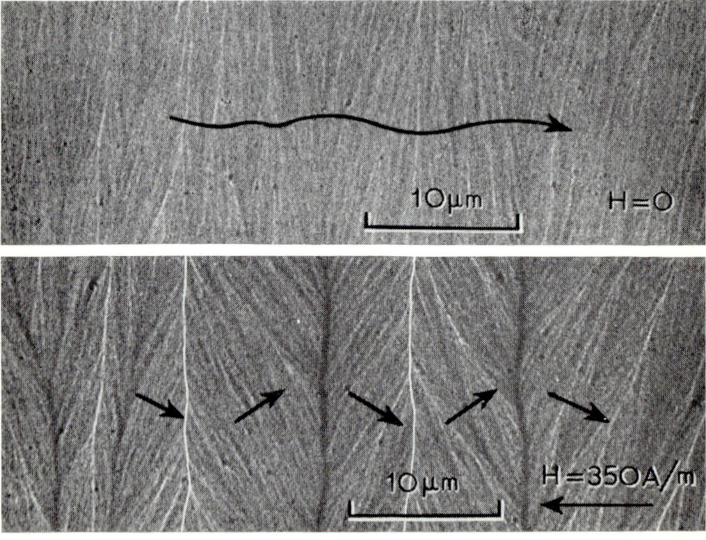

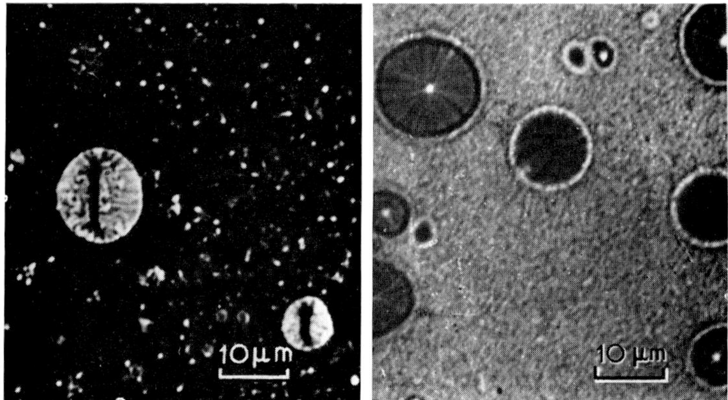

M. Domain structure of small circular particles of Permalloy. (*Cohen, 1965b*) *pages* 81, 87

N. Bitter pattern observed on a single crystal platelet of nickel cobalt 5000 Å thick. 112 μm $\times$ 35 μm showing the effect of a small applied field in displacing the domain boundaries. These crystals are prepared by the reduction of the metallic halide. (*Dr De Blois, private communication*) *pages* 5, 39, 47

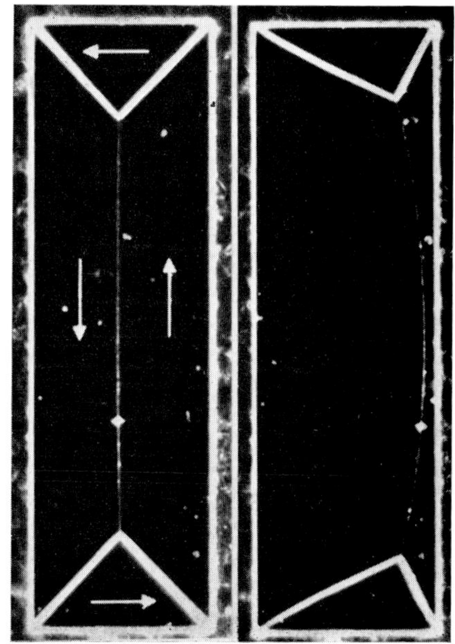

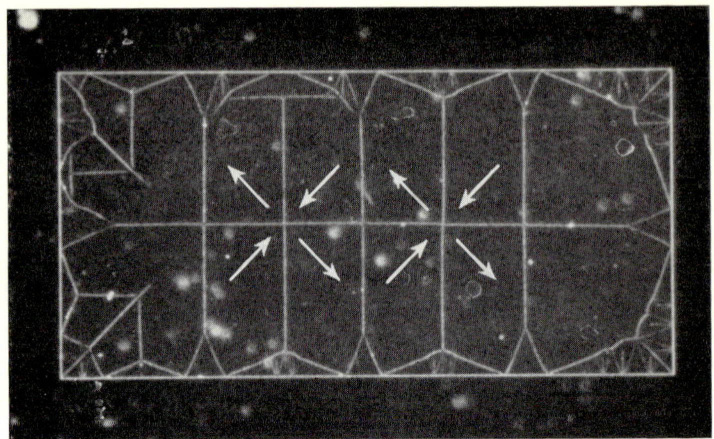

O. Bitter pattern observed on a single crystal platelet of Permalloy (86% Ni 14% Fe 100 μm $\times$ 200 μm) showing a domain structure resembling the Néel block with p-type closure domains (cf. Fig. 5.3). (*Dr De Blois, 1968, private communication*) *page* 55

P. Lorentz electron micrograph of domains in a single crystal iron film; the 'bend contours' are due to the bending of the specimen. ($\times$ 10,000) (*Dr D. C. Hothersall, private communication*) *pages* 9, 20, 21

domains associated with the missing spot are dark and the others associated with the remaining spot are light. By suitable adjustment of the aperture, domains with intermediate orientation can be shown up in grey rather than white or black.

The Lorentz method has the advantage that the physical structure of the specimen, stacking faults, dislocations and other crystalline defects, are imaged simultaneously with the domains. (The wealth of detail may in fact on occasion be an embarrassment, obscuring the fine detail of the domain configuration.) Other features of the image, which are associated with the wave properties of electrons are, for instance, the extinction fringes which are produced in wedge-shaped crystals and contour fringes which are associated with bending of the crystal (Plate P). It is natural to draw an analogy with the formation of Newton's rings and wedge fringes, but this is misleading; the rigorous theory of the formation of these fringes and of the images of crystalline defects involves a consideration of Schrödinger's equations applied to electron diffraction. However this is outside the scope of the present discussion which is concerned with the structure of domains rather than with the details of image formation.

The defocussing method may be looked upon from the aspect of phase contrast microscopy. The two antiparallel domains introduce a phase shift in the electron beam as it passes through the specimen, thus producing a deviation in the transmitted beam (eqn 2.2). In the converging condition this is in some respects similar to a Fresnel biprism of refractive index n introducing path differences of $\pm(n - 1)t$ and phase shifts of $\pm 2\pi(n - 1)t/\lambda$; this effect can be demonstrated by the production of interference fringes analogous to those produced optically with the Fresnel biprism (Plate G).

The phase shift is converted into an intensity difference by defocussing the objective. Similarly the displaced aperture (in focus) method may be looked upon as a Schlieren method of phase contrast. With this approach a fuller theory of the image formation can be developed (see Grundy and Tebble, 1968) but the treatment given above is sufficient for the present purposes.

It is necessary to differentiate clearly between the two classes into which specimens fall in transmission electron microscopy. (i) Bulk specimens which must be reduced in thickness to some 2000 Å to be transparent to 100 kV electrons. Metallic specimens are prepared by successive rolling or grinding, with intermediate annealing to remove strain, down to a few microns and then electropolished by one of the standard electropolishing solutions until some part of the foil is found by examination to be thin enough. Non-metallic specimens must be thinned by chemical etching. (ii) Specimens prepared by deposition either by evaporation, by electro-deposition or 'chemically' from solution are referred to as thin films and their properties are strongly dependent on the method of preparation. They are usually highly polycrystalline although single crystals may also be prepared in this way by epitaxial growth. Thin films are considered in Chapter 7.

2.5 Choice of methods

The Bitter method is so simple and convenient that it is still the most favoured where high resolution is not required, and most of the significant results in experimental domain work have been obtained by using this technique. Even when high resolution (better than 10^4 Å) is needed it is often useful to carry out preliminary observations by using the powder technique. It has the disadvantage that it can be used only at room temperature. Bulk specimens must be single crystals or at least contain a sizable crystallite so that the domain structure is sufficiently large to be observed. This last limitation does not apply to thin evaporated films with low anisotropy, as the direction of domain magnetization is controlled by the mode of evaporation rather than directly by the crystal structure, and the domain structure of polycrystalline evaporated films can be observed directly by the Bitter method. The resolution of the method has been increased by an order of magnitude by the Craik replica technique.

Magneto-optical methods may be used over a range of temperature but have about the same resolution as the Bitter technique, i.e. the optical limits of 10^4 Å. The speed of response to rapid changes in domain structure is very high and this method has been used to observe domain wall motion in rapidly changing applied fields.

The Lorentz method has by far the highest resolution – in the Fresnel method about 50 Å – but it is difficult to take full advantage of this because of the presence of the detailed background structure of the crystal lattice. There is the additional advantage that the specimen can be heated and subjected to magnetic fields and to strain whilst under observation. The crystalline orientation and crystal structure can be determined at the same time from the diffraction pattern in the electron microscope. This method also has the advantage that the area of the specimen within the field of view may be quite small (a few microns across) so that the domain structure of a single grain of a polycrystalline specimen can be observed and the method is not therefore limited to single-crystal specimens. There is, however, the serious limitation that the specimen must be transparent to the electron beam and must therefore be in the form of a thin foil or film about 2000 Å thick. Not only is the polishing of bulk specimens to this required thickness tedious and often difficult, but the final specimen may be significantly different in its physical crystalline structure from the bulk specimen. Evaporated films would seem to be eminently suited for this method but they are usually deposited on a substrate from which the film must be removed; this may alter completely the characteristics of the film.

The choice of a method to be used in a particular problem depends very largely upon the resolution which is required. It is obvious that if the very fine detailed structure of a particular domain configuration is to be investigated Lorentz electron microscopy has much to offer, but where optical resolution is sufficient the Bitter method is the most convenient.

3: Magnetic Fields and Magnetic Energy

3.1 Magnetic moment and 'free pole'

Before considering the magnetization process in detail it is necessary to give an outline of the background to certain aspects of magnetism and to define a number of parameters. Magnetic moment may be defined in terms of the expression for the magnetic field due to a dipole, moment M (Fig. 3.1a).

$$H_r = \frac{2M \cos \theta}{r^3}, \quad H_\theta = \frac{M \sin \theta}{r^3} \tag{3.1}$$

and the intensity of magnetization I is the magnetic moment per unit volume, i.e.

$$I = \frac{M}{V} \tag{3.2a}$$

The magnetic moment may be written $M = 2ml$, when m is, by analogy with the electric dipole, the pole strength of a magnetic dipole length $2l$. Referring to Fig. 3.1a, when the magnetized specimen has cross-section A and length $2l$, the intensity of magnetization can be written

$$I = \frac{M}{V} = \frac{m}{A} \tag{3.2b}$$

i.e. I is the magnetic pole per unit area of section; this 'free' magnetic pole may be considered as being concentrated at each end-face of the specimen with a surface pole density analagous to surface charge density $\sigma = I$. This device of using the free-pole concept is widely applied in magnetism generally and in the study of domains in particular. There is no suggestion that free poles 'exist'; the concept is simply one of mathematical convenience.*

* The spontaneous magnetization is written I_s, and in this work refers to the value at room temperature; a specimen although spontaneously magnetized in each domain may possess a relatively low overall magnetization, and only when magnetized to saturation by an applied field does the

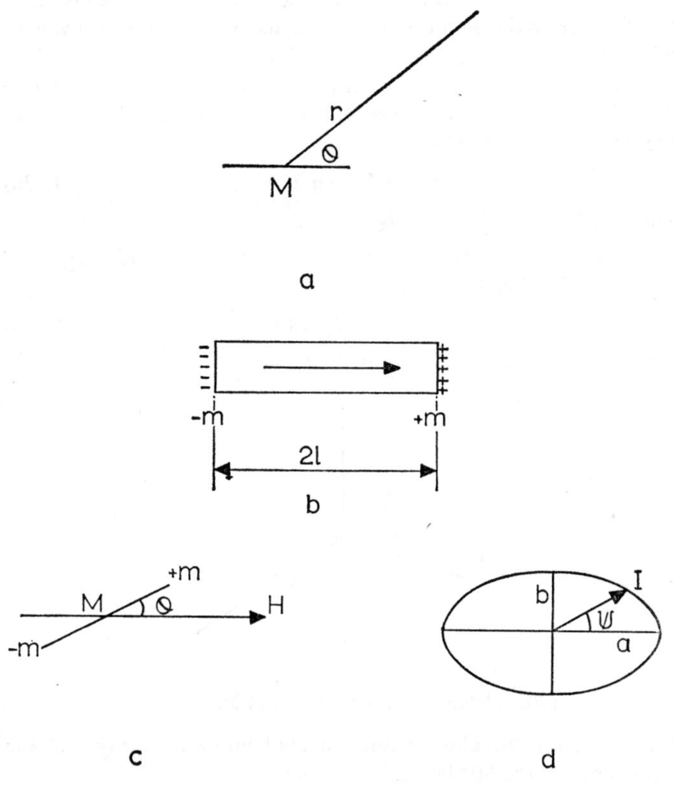

FIG. 3.1 (a) Field due to a magnetic dipole. (b) Representation of a magnetized object magnetic moment $M = 2\ ml$ with free pole M arising at the end-surfaces. (c) Magnetic moment $M = 2\ ml$ in a magnetic field H. (d) Magnetized ellipsoid with demagnetizing factors of N_a and N_b in directions a and b.

---

magnetization reach the spontaneous value. The saturation magnetization determined experimentally in this way is taken as equal to the spontaneous magnetization, and in this book the terms are assumed to be interchangeable.

The formation of 'free pole' at the end-surfaces may be considered as arising from the discontinuity in the magnetization from a value I within the specimen to zero outside; it is in fact the component of the magnetization normal to the surface which gives rise to the free pole. The end-surface of the specimen, as in Fig. 3.1b, is normal to the magnetization so that

$$\sigma = \mathbf{n}.\mathbf{I}_1 - \mathbf{n}.\mathbf{I}_2 = I, \qquad (3.3a)$$

since

$$I_2 = 0 \quad .$$

This free-pole effect may be produced at any surface and the general expression is (Fig. 3.2)

$$\sigma = \mathbf{n}.\mathbf{I}_1 - \mathbf{n}.\mathbf{I}_2 = I_1 \cos \alpha_1 - I_2 \cos \alpha_2. \qquad (3.3b)$$

Within a domain configuration, free poles may arise at a domain

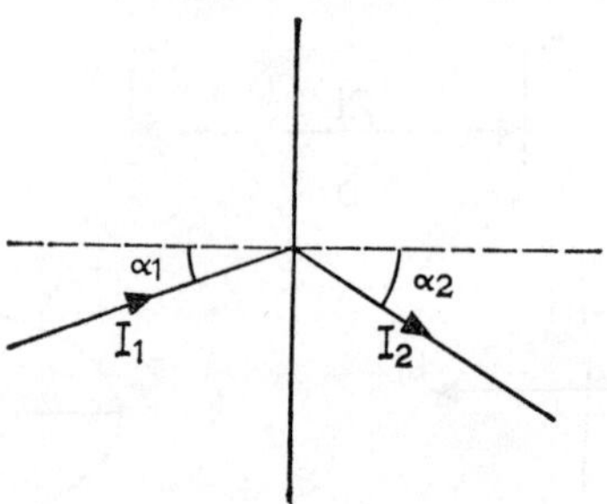

FIG. 3.2 Magnetization vectors at an interface.

boundary; for the closure domain configuration of Fig. 1.4 the expression for the 90° boundary becomes

$$I(\cos \alpha_1 - \cos \alpha_2) = \sigma \qquad (3.3b)$$

since $I_1 = I_2 = I$. At an end-closure domain in zero field the usual arrangement is $\alpha_1 = \alpha_2 = 45°$ so that $\sigma = 0$.

3.2 Demagnetizing fields and demagnetizing factors

There is associated with the surface free pole of Fig. 3.1b a magnetic field in the direction opposite to I, which close to the end-surface is $2\pi\sigma = 2\pi I$; this field falls off rapidly away from the ends and takes a mean value $H_D = -NI$; H_D is the demagnetizing field and N the demagnetizing factor. In a specimen of the shape shown, the demagnetizing field is not uniform and the magnetization also will be non-uniform; only in an ellipsoid are I and H_D uniform, but for many purposes this point is often conveniently, if unjustifiably, ignored.

For a sphere $N = 4\pi/3$; for the ellipsoid of revolution, polar semi-axis a, diameter $2b$, magnetized in a direction angle ψ with the a axis, (Fig. 3.1d)

$$N = N_a \cos^2 \psi + N_b \sin^2 \psi \tag{3.4a}$$

with

$$N_a + 2N_b = 4\pi \tag{3.4b}$$

The demagnetizing factor for an infinite cylinder ($a = \infty$) with elliptic cross-section and semi-axes b, c,

$$N_b = \frac{4\pi c}{b + c}, \qquad N =_c \frac{4\pi b}{b + c}. \tag{3.4c}$$

For a flat spheroid $b = c \gg a$

$$N_b = N_c = \frac{\pi^2 a}{b}. \tag{3.4d}$$

The following values of N_a for $q = a/b$ are for an ellipsoid of revolution (Fig. 3.1d; $b = c$)

q	0·1	0·5	1	2	5	10	20	50
$N_a/4\pi$	0·86	0·53	0·33	0·17	0·056	0·020	0·0067	0.0014

(It is important to emphasize that only for an ellipsoid is H_D uniform so that the use of $H_D = -N_D I$ can be justified.) The demagnetizing factor for a cube uniformly magnetized in any direction is equal to that for a sphere, i.e. $4\pi/3$; for cylinders and rods the values for ellipsoids give approximate values for N, again assuming uniform magnetization. This assumption can be justified if the uniaxial anisotropy of the material is very high so that the magnetization is always parallel to the easy direction. Further values of N for ellipsoids and rods together with references are given in Craik and Tebble (1965).

3.3 Magnetic induction, permeability, and susceptibility

The process of magnetization has been discussed in terms of the intensity of magnetization I; an alternative parameter is the magnetic induction B which is defined by the expression

$$B = H + 4\pi I. \tag{3.5a}$$

Where H is zero, B is clearly proportional to I and $B = 0$ when $I = 0$. For some applications where H is large, as in permanent magnets, this is no longer true and it is then necessary to specify whether the coercive field H refers to the state $I = 0$ or $B = 0$. The magnetic permeability is $\mu = B/H$ or $\mathrm{d}B/H$ and

$$\mu = 1 + 4\pi\kappa. \tag{3.5b}$$

The parameters associated with the magnetization curve may now be summarized. These are the coercivity H_c, the reverse field required to bring the magnetization I (or induction B) to zero; the remanence, i.e. the magnetization I or induction B at $H = 0$; the magnetic susceptibility κ at any point on the magnetization curve is I/H; the differential susceptibility $\mathrm{d}I/\mathrm{d}H$.

The reversible susceptibility κ_{rev} is measured by applying a sufficiently small change in magnetic field H giving $\kappa_{\mathrm{rev}} = \Delta I/\Delta H$ so that the change in magnetization is reversible (corresponding to 'a' in Fig. 1.2 inset). Experimental methods for measuring κ_{rev} usually make use of small alternating fields. In the demagnetized state κ_{rev} is the initial susceptibility κ_0; there are corresponding terms for the permeability μ, μ_{rev} and μ_0. A convenient method of demagnetizing a specimen is by the application of an alternating field of progressively reducing amplitude.

3.4 Magnetic energy

Any discussion of domain structure must involve a consideration of the expressions for the energy, and since the subject of energy as associated with magnetic fields (magnetic energy) is often ill-understood it is hoped that the following brief introduction will be helpful.

The internal energy per unit volume, E, of a magnetized body in a magnetic field H is given by the expression

$$\int \mathrm{d}E = \int \mathrm{d}Q + \int H\,\mathrm{d}I \tag{3.6}$$

$\mathrm{d}Q$ is the heat given to the body and $H\,\mathrm{d}I$ is the work done in changing the magnetization; I and H are taken for the sake of simplicity as being in the same direction. The term $H\,\mathrm{d}I$ corresponds to the term $(-p\,\mathrm{d}v)$ in the corresponding equation expressing the first law of thermodynamics. Where the intensity of magnetization is proportional to H (i.e. constant susceptibility $\kappa = I/H$)

$$\int_0^I H\,\mathrm{d}I = \tfrac{1}{2}HI. \qquad (3.7a)$$

This expression $\frac{1}{2}HI$ is sometimes quoted as the internal magnetic energy, but it is only true where I is proportional to H, as for instance with the demagnetizing field $H_D = -NI_D$ giving

$$E_D = -\tfrac{1}{2}NI^2. \qquad (3.7b)$$

In the expression for the total internal energy E' account is taken not only of the work done in magnetizing the object but also of its potential energy in the magnetic field. The potential energy of a magnetized object, moment $\mathbf{M}$, in the magnetic field is

$$-\mathbf{M}.\mathbf{H} = -MH\cos\theta, \qquad (3.8)$$

or in terms of energy per unit volume $-I.H$.
Then the expression for the total energy (enthalpy) becomes

$$E' = E - HI \qquad (3.9a)$$

$$\int \mathrm{d}E' = \int \mathrm{d}Q - \int I\,\mathrm{d}H, \qquad (3.9b)$$

where for convenience I and H are taken as parallel. The last term becomes $\int_0^H I\,\mathrm{d}H = \frac{1}{2}HI$ giving for the energy due to the demagnetizing field

$$E_D' = +\tfrac{1}{2}NI^2. \qquad (3.10)$$

The internal energy includes magnetocrystalline energy, magnetostriction and the free-pole energy arising at domain boundaries inclusions and imperfections. The problem of deriving the equilibrium conditions for a particular domain configuration is essentially one of expressing the internal energy E associated with these effects and obtaining the condition for minimum total energy $E' = E - HI$. It is necessary therefore to discuss the processes involved and to give the appropriate expressions for the energy.

3.5 Free-pole energy

The energy associated with demagnetizing fields and with free-pole effects in general is of particular significance in the treatment of domain structure. As stated above, the demagnetizing field $H_D = -NI$ is, near the end of a specimen $-2\pi I = -2\pi\sigma$, where σ is the surface density of the free pole. The energy is therefore $\pi I^2 = \pi\sigma^2$ and, with $I \sim 10^3$, takes the value of $\sim 10^6$ to 10^7 erg cm^{-3}; this is for many materials greater than any other factor such as the magnetocrystalline energy (see section 3.6). This free-pole effect will also arise at any interface across which there is a discontinuity in the normal component of the magnetization. In Fig. 3.2

$$\sigma = I(\cos\theta_1 - \cos\theta_2). \tag{3.11}$$

Even though σ may be considerably less than the value given above, the contribution to the energy of the system will be significant even for quite small differences in θ in eqn (3.11).

The result is that in virtually all domain structures the boundaries configurations are those which produce minimum free-pole energy; either by reducing the magnitude of the free pole or by 'spreading out' the free pole over a large area so as to reduce σ, the free-pole surface density. Any alternation in the sign of the free pole over a given surface will, of course, also reduce the energy of the system.

3.6 Magnetocrystalline anisotropy

In most ferromagnetic materials there are preferred directions of magnetization, usually along one or other of the major crystalline axis. Magnetocrystalline anisotropy arises from the coupling between the electron spin and the orbital motion of the electron. The directional properties of the electron orbit are highly oriented with respect to the crystal lattice; thus the effect of an applied magnetic field on the spin moments depends on the spin orientation with respect to the crystal lattice. In hexagonal cobalt, for instance, the easy direction is parallel to the c axis and a finite field is required to rotate the magnetization out of the preferred direction (Fig. 3.3.). The energy associated with this process is

$$E_K = K_1 \sin^2\theta + K_2 \sin^4\theta + \ldots \tag{3.12}$$

where θ is the angle between the preferred direction and the direction of magnetization. For most purposes it is sufficient to write

$$E_K = K \sin^2 \theta \tag{3.13}$$

where K is the magnetocrystalline anisotropy constant. For cobalt $K = +5{\cdot}3 \times 10^6$ erg cm^{-3}. Where K is negative the easy direction is perpendicular to the c axis, as in cobalt at high temperatures (section 4.2). It is convenient to consider the constant K in terms of the equivalent magnetic field, that is the field which is required to rotate the magnetization against the anisotropy. This is given by

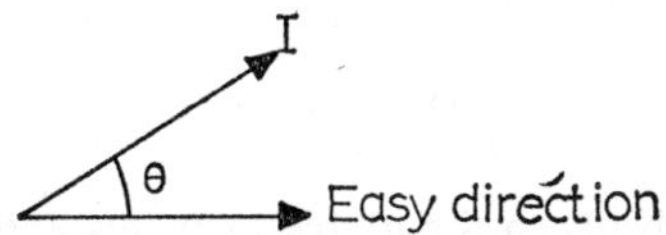

FIG. 3.3 Direction of magnetization with respect to the easy direction.

$H_K = 2K/I_s$ giving for cobalt $H_K = 7400$ oersted; for MnBi $K = 8{\cdot}9 \times 10^6$ erg cm^{-3}, $H_K = 40{,}000$ oersted.

In a material with a cubic crystalline structure such as iron the expression for E_K, neglecting higher terms, is,

$$E_K = K(\alpha_1^2 \alpha_2^2 + \alpha_1^2 \alpha_3^2 + \alpha_2^2 \alpha_3^2) \tag{3.14}$$

where α_1, α_2 and α_3 are the direction cosines of the magnetization vector with respect to the cube axis. Where K is positive (as in iron) the preferred directions are parallel to the $\langle 100 \rangle$ cube axes, and for K negative (as in nickel) are along the $\langle 111 \rangle$ cube diagonals. Values of K and H_K are:

Iron $K = 4{\cdot}8 \times 10^5$ erg cm^{-3}, $H_K = 560$ oersted.
Nickel $K = -4{\cdot}5 \times 10^4$ erg cm^{-3}, $H_K = 185$ oersted.

In general, materials with a uniaxial structure have higher anisotropy constants than those with cubic symmetry.

3.7 Induced anisotropy

A preferred direction of magnetization may be induced in a specimen by annealing in a magnetic field. This is of particular importance in alloys and is particularly noticeable in the nickel-iron

(Permalloy) alloys. It may be explained in terms of the direction ordering of atom pairs. Instead of a random array of A and B atoms, each carrying a magnetic moment, AA, AB and BB pairs may be aligned in a particular direction with respect to the applied field so as to reduce the magneto-elastic energy. A corresponding alignment may occur with vacancies. This introduces an anisotropy in the magnetostatic field associated with the atomic moments compared with a random arrangement which is isotropic.

Considering the expression for the energy due to the magnetostatic interaction between a pair of parallel dipoles of moment μ and separation r, we have

$$\frac{\mu^2}{r^2}(1 - 3\cos^2\phi) \tag{3.15}$$

with ϕ the angle between the dipole moments and the line joining their centres. In a simple cubic array of parallel dipoles, the total energy of a dipole due to interaction with its six nearest neighbours is the sum of an equal number of terms in $\sin^2\phi$ and $\cos^2\phi$ and is thus independent of ϕ as $(\sin^2\phi + \cos^2\phi) = 1$; a single impurity or vacancy will not alter the symmetry. However, the introduction of a pair of impurities or vacancies on neighbouring sites will result in an anisotropy with terms proportional to $\sin^2\phi$, so that this source of anisotropy can be written in a form similar to that for magnetocrystalline anisotropy $K\sin^2\phi$.

3.8 Magnetostriction

A change in magnetization in a ferromagnetic crystal is usually accompanied by a change in dimensions – the magnetostriction effect. The magnetostriction coefficient λ_s is defined as the fractional change in length δl associated with the change in magnetization I from zero to saturation. The size and sign of the effect depend critically upon the direction of δl and I with respect to the crystal axis, and the expressions needed to give a complete description of the magnetostriction energy are involved and cumbersome. It is sometimes justifiable to use a mean value for the magnetostriction constant λ_s assuming the material to be isotropic, but the only real justification in most cases is that the procedure greatly simplifies the energy expressions.

With the application of a stress to a material the magnetostrictive effect is associated with the introduction of preferential axes of magnetization similar to the effect of magnetocrystalline anisotropy. Where λ_s is positive, for instance, a longitudinal stress produces a preferred direction in the direction of positive tension. An example of this effect is seen in a stretched iron wire where, for a high value of tension, λ_s is positive and the specimen has many of the properties of a uniaxial crystal such as cobalt, with a preferred axis of magnetization in the direction of the tension, i.e. along the axis of the wire. A compressive stress rotates the easy axis of magnetization so that it is at right angles to the direction of stress (cf. Plate C). The effect of stress σ upon a ferromagnetic material can be allowed for by adding to the coefficient K in eqn (3.13) a term $K_\sigma = \frac{3}{2}\lambda_s\sigma$ to give

$$(K + \tfrac{3}{2}\lambda_s\sigma) = K + K_\sigma. \qquad (3.16)$$

4: The Structure of the Magnetic Domain

4.1 Domain boundaries—Bloch walls

The expression of the exchange interaction field in terms of an overall internal magnetic field proportional to the magnetization has been most useful in the historical development of the theory of ferromagnetism, and the terms Weiss field and Weiss constant are still used as a matter of convenience. However, in the study of the structure of domain walls the interaction must be considered in more detail as an interatomic phenomenon affecting the orientation of the individual spin moments across the boundary.

As described in the introduction, the structure of the domain wall

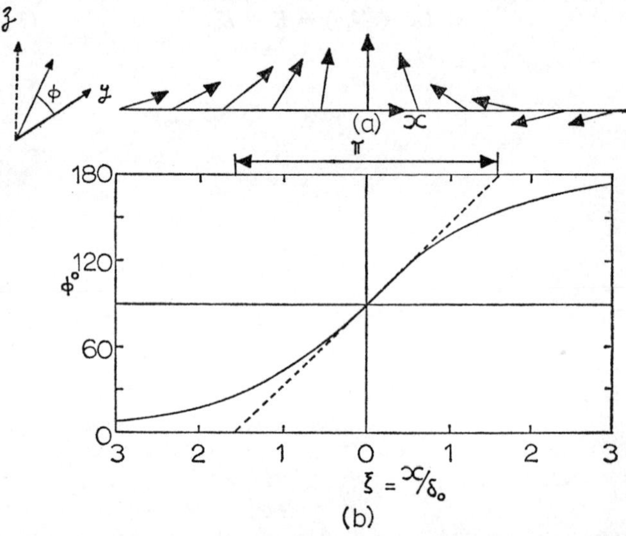

FIG. 4.1 (a) Variation in orientation ϕ of the magnetization vectors in a domain boundary. (b) Curve of ϕ against reduced distance from the centre of the wall.

represents a balance between the exchange effects tending to produce a very thin wall and anisotropy forces tending to produce a very wide wall. In Fig. 4.1a the exchange interaction energy between neighbouring atoms can be expressed as

$$-2J\,\Sigma\,\mathbf{S}_i\mathbf{S}_j = -2JS^2\,\Sigma\cos\phi_{ij} \tag{4.1a}$$

with $\mathbf{S}_i\mathbf{S}_j$ the spin vectors on neighbouring atoms oriented at an angle ϕ_{ij} with each other; J is the exchange integral, which for ferromagnetic materials is positive so that the energy is a minimum when $\mathbf{S}_i$ and $\mathbf{S}_j$ are parallel. For the boundary problem, we are interested only in terms involving ϕ in eqn (4.1a) which becomes for ϕ small

$$JS^2\,\Sigma\,\phi_{ij}^2. \tag{4.1b}$$

The simple boundary configuration of Fig. 4.1a is 'one-dimensional' in that ϕ varies only in the direction x so that with an angle between neighbouring atoms of interatomic distance a_0, $\phi_j = a_0(\mathrm{d}\phi/\mathrm{d}x)$ and the equation becomes

$$JS^2\left(\frac{\mathrm{d}\phi}{\mathrm{d}x}\right)^2 a_0^2. \tag{4.1c}$$

The energy per unit volume is then, for a simple cubic arrangement of atoms

$$E_{\mathrm{ex}} = \frac{JS^2}{a_0^3}\left[\left(\frac{\mathrm{d}\phi}{\mathrm{d}x}\right)a_0\right]^2 = \frac{JS^2}{a_0}\left(\frac{\mathrm{d}\phi}{\mathrm{d}x}\right)^2 = A\left(\frac{\mathrm{d}\phi}{\mathrm{d}x}\right)^2 \tag{4.2}$$

where A is the exchange constant and is of the order of 10^{-6} erg cm^{-1} (see Table 1.1).

The expression for the energy of the boundary as a whole involves not only the exchange energy but the magnetocrystalline anisotropy energy which for uniaxial anisotropy is $K\sin^2\phi$ and thus

$$\gamma = \int_{-\infty}^{+\infty}\left[A\left(\frac{\mathrm{d}\phi}{\mathrm{d}x}\right)^2 + K\sin^2\phi\right]\mathrm{d}x \tag{4.3}$$

per unit area. The exchange contribution would be reduced for $(\mathrm{d}\phi/\mathrm{d}x)$ small, i.e. for a boundary of infinite width: the anisotropy term is a minimum for 'zero width' with the magnetization parallel and antiparallel to the easy direction in adjacent domains and $\phi = 0$ or $180°$ (Fig. 1.3d). The boundary configuration thus represents a balance between the exchange energy and the anisotropy energy. The total energy given by eqn (4.3) is a minimum if $A(\mathrm{d}\phi/\mathrm{d}x)^2 = K\sin^2\phi$, i.e.

$$\frac{\mathrm{d}\phi}{\mathrm{d}x} = \left(\frac{K}{A}\right)^{\frac{1}{2}} \sin\phi. \qquad (4.4a)$$

This gives

$$\frac{x}{\delta_0} = \ln\tan\tfrac{1}{2}\phi \qquad (4.4b)$$

$$\cos\phi = -\tanh\left(\frac{x}{\delta_0}\right)$$

$$\sin\phi = \mathrm{sech}\left(\frac{x}{\delta_0}\right);$$

it is convenient to use $\delta_0 = (A/K)^{\frac{1}{2}}$ and $\gamma_0 = (AK)^{\frac{1}{2}}$ as parameters of length and energy. Substituting for ϕ in eqn (4.3) gives for the boundary energy per unit area ('surface tension')

$$\gamma = 4\gamma_0 = 4(AK)^{\frac{1}{2}}. \qquad (4.5)$$

The orientation, ϕ, of the magnetization vector across the boundary is shown in Fig. 4.1b; the 'width' of the boundary is obviously 'infinite' but an effective width δ is shown in Fig. 4.1b.

$$\delta = \pi\delta_0 = \pi\left(\frac{A}{K}\right)^{\frac{1}{2}}. \qquad (4.6)$$

For cobalt $\delta_0 = 50$ Å, giving $\delta \simeq 150$ Å; also $\gamma_0 = 2{\cdot}05$ erg cm^{-2} giving $\gamma = 8{\cdot}2$ erg cm^{-2} from eqn. (4.5).

These expressions apply only to a material with uniaxial anisotropy such as cobalt. Forms of anisotropy other than magnetocrystalline may be present, and in that event the treatment requires modification; clearly in a material with a very low value of K, strain anisotropy or induced anisotropy might be the effective factor.

For cubic materials the values of δ and γ depend on the boundary orientation, but for the configuration of Fig. 1.4d, e values for iron are

90° boundary $\delta = 520$ Å, $\gamma = 1{\cdot}07$ erg cm^{-2}
180° boundary $\delta = 1400$ Å, $\gamma = 1{\cdot}25$ erg cm^{-3}

allowing for magnetostriction effects. The orientation of boundaries in nickel is rather more complex but values for 180° (110) walls are $\delta = 2000$ Å, $\gamma = 0{\cdot}3$ erg cm^{-3}.

It will be noted that only the anisotropy and exchange energy are accounted for in this treatment, and the magnetostatic energy associated with the free-pole formation at the intersection of the wall with the surface of the specimen is neglected. This is justifiable in thick specimens when the width of the wall is small compared with the thickness of the specimen and the area of free pole is relatively small. In very thin films, however, the width of the wall is at some stage greater than the film thickness and it is energetically more favourable for the magnetization vectors to rotate in the plane of the film as in Fig. 7.1; this configuration is known as a Néel wall. The formation of Néel walls is discussed in the section on thin films (Chapter 7).

4.2 Simple domain structures

Expressions for the dimensions of a domain for the simple configuration shown in Fig. 1.4 can now be derived. Most of the theory of domain structures so far developed concerns single crystals. The subject of polycrystalline specimens is discussed in section 4.3. It is sufficient to say that the domain structures described in this section occur in single crystals ranging in linear dimensions from centimetres down to microns. For a specimen with overall magnetization zero, subdivided as shown in Fig. 1.4b, the demagnetizing energy per unit volume $\frac{1}{2}NI^2$ decreases with increasing subdivision; N is inversely proportional to the number of domains n and is given by

$$N = \frac{3{\cdot}4}{n}\cdot\frac{a}{L} \tag{4.7}$$

where $\dfrac{a}{L} \ll 1$ and obviously the demagnetizing factor decreases with the length of the specimen. In the derivation of this expression which is tedious rather than difficult, it is necessary to assume that the magnetization in each domain is uniform. The subdivision into domains is limited by the increase in the energy associated with the domain walls; with the number of walls large $n \sim n - 1$; each wall is of area aL and energy γ per unit area, and the total boundary energy is $nLa\gamma$.

The energy for the domain configuration as a whole is then

$$W = 1{\cdot}7I_s^2\left(\frac{a}{nL}\right)a^2L + n\gamma aL$$

$$= 1{\cdot}7I_s^2\frac{a^3}{n} + n\gamma aL. \tag{4.8}$$

This is a minimum where $\mathrm{d}W/\mathrm{d}n = 0$, i.e. when

$$n = \left(\frac{1{\cdot}7I^2a^2}{\gamma L}\right)^{\frac{1}{2}} \tag{4.9a}$$

and the domain spacing d is

$$d = \frac{a}{n} = \left(\frac{\gamma L}{1{\cdot}7I^2}\right)^{\frac{1}{2}}. \tag{4.9b}$$

Thus for a single crystal 1 cm in length, $\gamma = 1$ erg cm^{-2} and $I = 1700$ e.m.u. cm^{-3}, eqn. (4.9b) gives $d = 4{\cdot}5 \times 10^{-4}$ cm.

The derivation of the corresponding expression for the checker-board structure and the honeycomb structures of Figs. 1.4c and 1.4g are similar giving

$$d = \left(\frac{2\gamma L}{I_s^2}\right)^{\frac{1}{2}} \quad \text{and} \quad d = \left(\frac{4{\cdot}25\gamma L}{I_s^2}\right)^{\frac{1}{2}} \tag{4.10}$$

respectively.

The essential point of these domain configurations is that they reduce the overall free-pole energy by the alternation in the sign of the free pole. This is made more effective either by reducing the 'wavelength' of the alternation or by the interlocking of the free-pole pattern on the end-surface. This is illustrated in Plates B and D which show the domain pattern observed on the basal plane of high anisotropy materials, magnetized normal to this plane. These patterns represent a cross-section of the domain structure (cf. Fig. 1.4c, g). The formation of reverse spikes also has the same effect. The growth of these spikes into reverse domain provides a mechanism for reversing or reducing the overall magnetization of the specimen (Fig. 1.4f). Reverse spikes may also be formed at inclusions or grain boundaries where free pole is produced (Fig. 4.3); again the growth of these spikes provides a means of reversing the magnetization (Plate I).

Where a closure structure of the type shown in Fig. 1.4d is formed there is in the demagnetized state no free-pole contribution to the energy. The magnetization in the closure domain is parallel to the surface at the ends of the specimen so that no free pole arises (Plate IV). Across the 90° boundary of the closure domain the normal component of the magnetization is continuous (equal to $I_s \cos 45°$) and this surface is also free of demagnetizing effects. In a uniaxial crystal, however, with the easy axis parallel to the length L there is a contribution from the magnetocrystalline anisotropy; the energy per unit volume $K \sin^2 \phi = K \sin^2 \pi/2 = K$ giving for a total volume $2nad^2/4$ an energy $Knad^2/2 = Ka^3/2n$.

The energy of the system is then, neglecting the small contribution to the boundary energy due to the closure domains,

$$W = \frac{Ka^3}{2n} + n\gamma aL \tag{4.11a}$$

giving for the minimum energy

$$n = \left(\frac{Ka^2}{2L\gamma}\right)^{\frac{1}{2}}; \quad d = \frac{a}{n} = \left(\frac{2L\gamma}{K}\right)^{\frac{1}{2}}. \tag{4.11b}$$

In fact this type of structure is unlikely to be formed in a uniaxial material because of the relatively high energy. It is observed in cubic materials where the magnetization in both the closure domains and the main domains can lie in easy directions so that the anisotropy energy is zero. There is, however, a magnetostrictive contribution corresponding to the term K in eqn (4.11) which is replaced by $\frac{1}{2}C_{11}\lambda_{100}^2$ where C_{11} is the elastic modulus and λ_{100} the magnetostriction constant in the [100] direction. These 90° closure structures are formed in all cubic materials and with a wide range of dimensions of from 1 mm in width down to 1 micron.

A closure structure of rather different type is formed in cubic crystals when the surface diverges a few degrees from planes containing the easy axis. The 180° boundaries which are observed on such surfaces are found to be decorated by fir-tree patterns as shown in Plate A. As shown in Fig. 4.2, these patterns are associated with

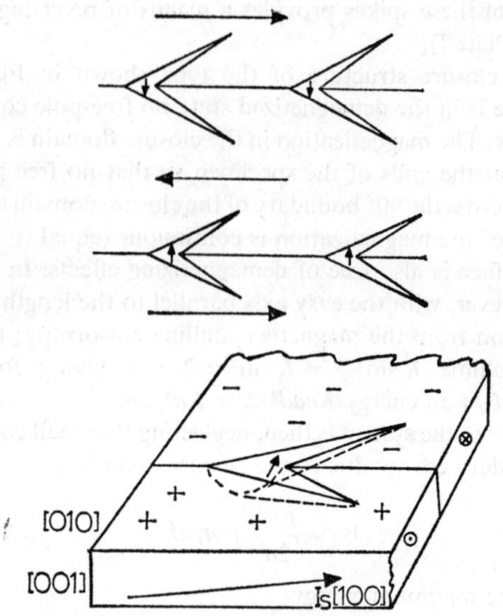

FIG. 4.2 Analysis of the fir-tree pattern of Plate H showing the direction of domain magnetization.

a closure structure which forms in such a way as to reduce the effect of the free pole arising from the slight misorientation of the specimen surface, away from (100). It will be seen that the fir-tree domain has a finite depth with domain walls within the specimen; these walls lie at approximately 45° with the main magnetization vectors satisfying as far as possible the condition for minimum free-pole density.

Closure structures are also formed at inclusions and other imperfections in the crystal structure – for instance at grain boundaries. Examples of the subsidiary closure structure at inclusions are shown in Fig. 4.3 (c–e). Although the 'spikes' formed on the inclusion do not make an angle of exactly 45° with the domain magnetization (this is obviously not geometrically possible), the deviation is small and the free-pole density relatively low; the structure is known as a Néel spike.

The domain structures described so far in this chapter are geo-

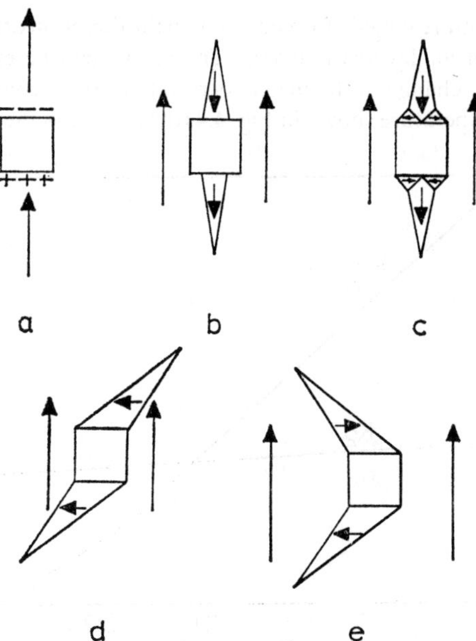

FIG. 4.3 (a) Free pole arising at an inclusion. Reduction of free-pole energy by the formation of (b) and (c) reverse spikes of magnetization, and of (d) and (e) Néel spikes.

metrically simple and consist essentially of 180° domains, i.e. domains magnetized parallel and antiparallel to a preferred axis with modification due to closure domain formation at the ends. The significance of the preferred axis is demonstrated in Plate H. These are Lorentz electron micrographs of electropolished cobalt foil taken over a range of temperatures. It will be seen that the structure consists at room temperature of the usual pattern of 180° boundaries parallel to the *c* axis. As the temperature is increased, this pattern is replaced by one in which the magnetization is progressively rotated until there are four easy directions of magnetization at about 270°C. The formation of closure structures at A and B is due to the presence of more than one easy axis of magnetization; they are not formed when cobalt is magnetically uniaxial. At 330°C a 180°

domain structure is again formed but with the direction of magnetization at right angles to the c axis. The results can be explained in terms of the change in the magnetocrystalline anisotropy constants with the temperature shown in Fig. 4.4 (the change in the magnetiza-

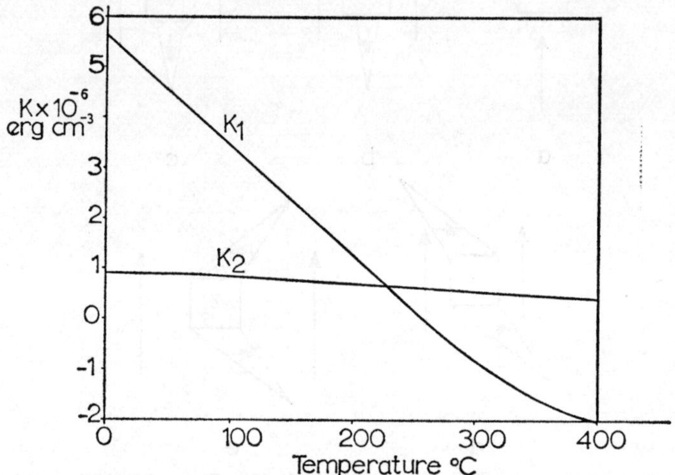

FIG. 4.4 Variation in magnetocrystalline anisotropy constants of cobalt K_1 and K_2 with temperature.

tion is not significant). The anisotropy energy expression from eqn (3.12) is

$$E_K = K_1 \sin^2 \phi + K_2 \sin^4 \phi.$$

At room temperature K_1 and K_2 are both positive, giving uniaxial magnetic properties with the [0001] hexagonal c axis an easy direction of magnetization. At 330°C K_1 is negative and larger than K_2; as will be seen below, K_1 is sufficiently large and negative for the easy direction to be at right-angles to the original easy c axis. The easy direction at intermediate temperatures is found by obtaining the energy minimum of the above equation, i.e. when

$$\frac{dE_K}{d\phi} = 0 \quad \text{and} \quad \frac{d_2E_K}{d\phi^2} > 0.$$

Above about 250°C where K_1 is negative and K_2 positive ϕ is given by

$$\sin^2 \phi = \frac{-K_1}{2K_2} \tag{4.12}$$

for the condition $0 \langle -K_1/2K_2 \rangle 1$. Thus when $-K_1 = K_2$ (at about 300°C) $\sin \phi = 1/\sqrt{2}$ giving for ϕ the values $\pi/4$ or $3\pi/4$, and thus there are two easy axes in the crystal.

4.3 Polycrystalline specimens

The domain configurations so far considered have been those formed in single crystals; with a uniform crystalline orientation throughout the specimen the preferred axes of magnetization are in the same direction throughout the crystal. Ideally the magnetization throughout a domain would be uniform, but as already mentioned there are deviations from this near the surface or near imperfections in the crystal.*

In a polycrystalline specimen there is normally a change in crystalline orientation and hence in the preferred directions of magnetization between neighbouring crystallites or grains. In some cases this is observed as a sharp change in direction of the boundaries across the grain boundary; it may also be accompanied by the formation of spike domains at the grain boundaries (Fig. 4.5 and Plate I). The domain structure within a single grain seems to follow the pattern of domain structures observed in large single crystals; this is so, not only in the large crystallites present in, say, silicon iron but also in the small grains about 1 micron across of less highly oriented materials as observed by electron microscopy. It is only when the grain size approaches that of a single domain particle that any

* Near the surface of a specimen the effect of the localized demagnetizing fields (arising from the 'free-pole' effect) is to rotate the magnetization away from the easy direction and there may be some modification of the domain structure near the ends of the specimen. This may require a correction to the spacing of the domains given by the expressions above. The net result depends on the relative magnitudes of terms involving the demagnetizing energy $2\pi I_s^2$ and the anisotropy constant K. The effect is often expressed in terms of a 'rotational permeability' $\mu^* = 1 + 2\pi I_s^2/K$ and is referred to in the literature as the 'μ^* effect'.

radical alteration is to be expected, and observation on this scale has been limited.

The presence of the grain boundary has thus much less effect on the domain structure than might have been expected; even the movement of a domain boundary under the action of an applied

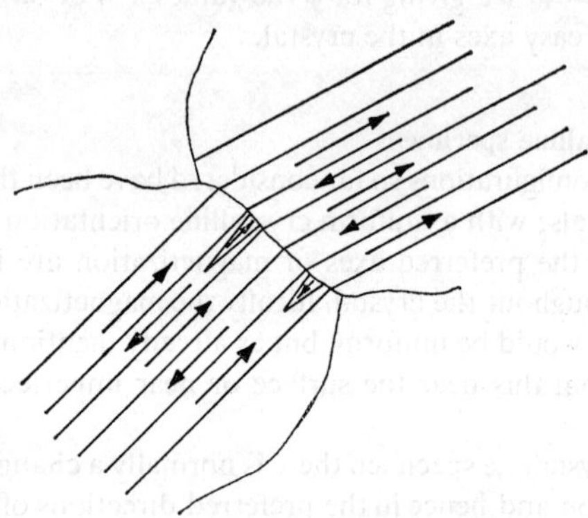

FIG. 4.5a Representation of domain structure at a grain boundary in polycrystalline silicon steel. The domain walls lie parallel to the easy axis in each grain. Note the spikes of reverse magnetization (cf. Plate H).

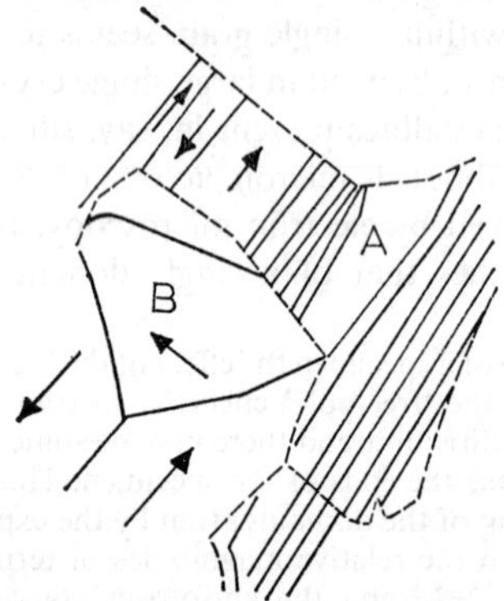

FIG. 4.5b Drawing of domain structure in cementite, in iron; the grain boundary between the cementite A and iron B is shown - - -.

field is not always delayed or impeded in any obvious way. (It could be said that this is typical of domain investigations.) The interaction between the domain boundary and the physical defect of the crystal lattice, such as dislocations and grain boundaries, seems to be quite unpredictable. The problem might be approached by allowing for the interaction between individual crystal grains, but at the moment there does not seem to be any way of doing this.

When the grain boundary is one between two phases there is some sort of continuity between the different domain configurations in the two regions, but this is not complete. Fig. 4.5b, for instance, shows the domain configuration in body-centred cubic iron containing particles of cementite Fe_3C which has an orthorhombic structure. The orthorhombic cementite has a high uniaxial magnetocrystalline anisotropy of the same order of magnitude as cobalt with the easy axis parallel to [001]. The domain structure in the iron (B) is similar to that of Fig. 1.4d with two easy directions of magnetization; the cementite (A) has a uniaxial domain structure similar to that of cobalt, consisting of 180° domains. It is obvious that this configuration must lead to a relatively high coercivity due to the 'misfit' between the two types of domain structure giving rise to free-pole effects. A similar effect is observed in cobalt when both the fcc and the hcp phases are present (see Craik and Tebble, 1965).

5: Domain Structures and the Magnetization Curves

The interest in the study of domains does not lie in simply analysing a static domain configuration to derive the conditions from minimum energy but is concerned with the effect of applying a magnetic field; it is in this way that an understanding of the magnetization

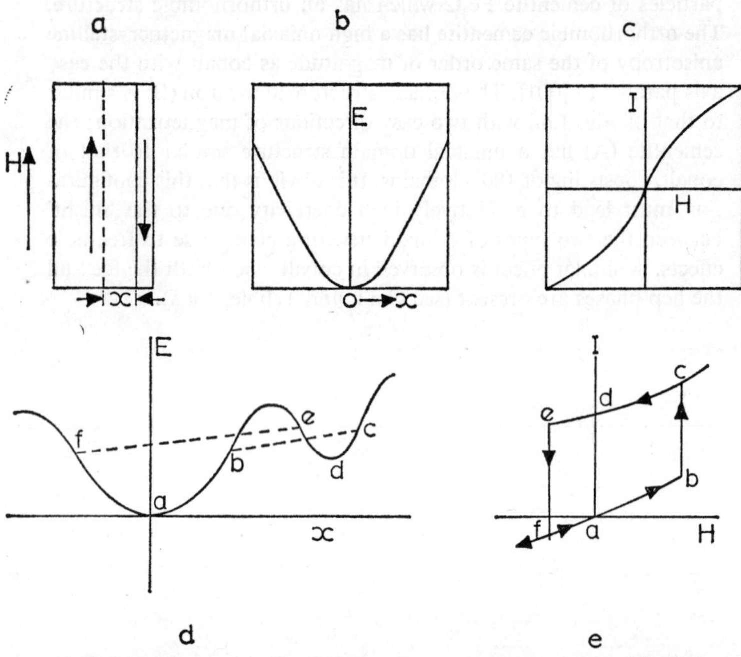

FIG. 5.1 (a) Movement of a 180° boundary across a specimen with (b) the corresponding energy (E, X) and (c) magnetization (I, H) curves. (d) Representation of the energy curve for a 180° boundary impeded by imperfections and (e) the corresponding magnetization (I, H) curve.

process and hysteresis is to be obtained. Consider the simplest configuration consisting of two antiparallel domains (Fig. 5.1a). The demagnetizing energy E_D is roughly equal to one half that of the uniformly magnetized specimen. Upon the application of a field in the z direction the domain wall is displaced, (x) so that the volume of the domain with I parallel to H increases at the expense of the antiparallel domain; the energy due to the demagnetizing field increases with x and obviously would eventually take the value appropriate to the uniformly magnetized crystal. This is shown diagrammatically in Fig. 5.1b. The displacement of the 180° domain boundary in a specimen with a closure structure is shown in Plate N (c.f. Fig. 1.4d and 6.1b). The actual form of the $E, x)$ relation is not straightforward but in any event it is not material at this point. The total energy of the system, per unit volume in an applied field, is thus

$$E' = E - HI = E - 2HI_s x/d \tag{5.1a}$$

where d is the width of the specimen. The boundary takes up the position given by $\mathrm{d}E'/\mathrm{d}x = 0$, that is

$$\frac{\mathrm{d}E}{\mathrm{d}x} = 2HI_s/d. \tag{5.1b}$$

The magnetizing curve is of the form shown in Fig. 5.1b. With

$$H = \left(\frac{\mathrm{d}E/\mathrm{d}x}{2I_s/d}\right) \quad \text{and} \quad I = 2I_s x/d. \tag{5.1c}$$

Each point on the (E, x) curve corresponds to a point on the (I, H) curve.

The magnetization curve in this example is reversible and clearly some modification of the picture is required to account for the shape of the hysteresis curve. It is reasonable to assume that as the boundary moves across the specimen it is impeded by impurities and by imperfections in the crystal and this is represented on the (E, x) curve by additional maxima and minima as shown (Fig. 5.1d). Now on applying a small field the boundary moves from a to an equilibrium position given by eqn (5.1a), but at some point b, $(\mathrm{d}E/dx)$ begins to decrease so that H is always greater than $(\mathrm{d}E/\mathrm{d}x)/(2I_s/\mathrm{d})$, i.e. H is too large for equilibrium. The system is now unstable and the boundary moves irreversibly and discontinuously to a new position of equilibrium c with a slightly larger value of $(\mathrm{d}E/\mathrm{d}x)$ which satisfies eqn (5.1b). The corresponding points a, b and c on the magnetization curve are shown in Fig. 5.1e with ab the reversible continuous part of the boundary movement and bc the discontinuous part corresponding to a Barkhausen discontinuity. With a further increase in field the position of equilibrium moves up the (E, x) curve beyond c. If H is reduced then the path taken will be c to d at which point the field is zero $(\mathrm{d}E/dx = 0)$. To return to the point a the field must be reversed at d and at e a discontinuous change in the reverse direction will take place; the boundary will then have moved to a new position of equilibrium f on the other side of the demagnetized position. A further increase in the field in the reverse direction will produce a further motion beyond f. On reducing the reversed field to zero the point a will be regained. In this way a small hysteresis loop is traversed and for this particular cycle a value for the coercivity can be derived. Note that if the reverse field had been increased beyond f to produce a further discontinuity then the miniature loop would not have passed through a.

The complete hysteresis loop would require a summation of these individual processes for the whole of the cycle and the coercivity for the specimen is a mean of the coercivities for the various processes. It is thus necessary to investigate the details of the microscopic processes which are involved in the delay in the domain wall motion.

5.1 Coercivity and the magnetization curve

Two simple examples will be given to demonstrate the nature of the problem of deriving expressions for the coercivity and to give some indication of the methods. It has already been indicated that the motion of the domain wall is held up by inclusions and imperfections in the crystal structure of the specimens. The two mechanisms to be considered are those associated with the delay in the motion of a domain wall at a non-magnetic inclusion, (i) with and (ii) without the formation of the Néel tie structure; these are illustrated in Fig. 5.2.

It should be emphasized that not only are the suggested models much simplified but the mathematical treatment is itself often elementary, to say the least. This is necessary because even with the most elementary model the mathematical expressions for the magneto-static field are often cumbersome and it is sometimes necessary for quite far-reaching simplifications to be made.

The Néel tie structure is shown in Fig. 5.2(i). It will be seen that

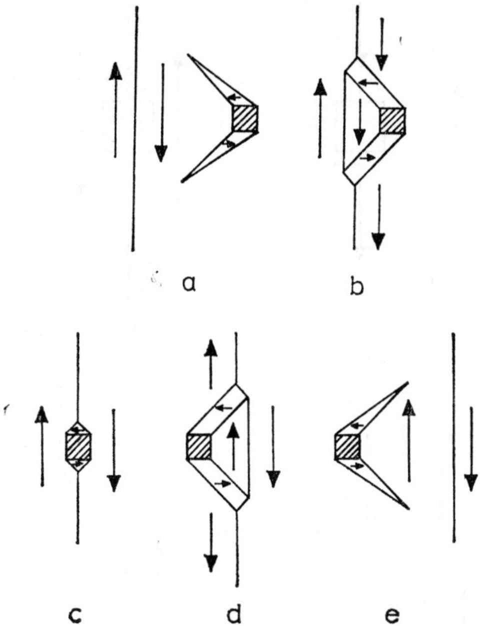

FIG. 5.2 (i) Movement of a 180° boundary, across a non-magnetic inclusion interacting with the Néel 'spike' structure (a, e), to form a 'tie' structure (b, d).

at an inclusion a subsidiary spike structure (a Néel spike) is formed which links up with the 180° boundary shown. As the 180° boundary moves away from the inclusion its motion is impeded by the closure structure which ties it to the inclusion. The elementary analysis of the energy of the system is simply one of balancing the magnetic term in HI with the change in wall energy as the tie structure is extended. For a cubic inclusion, side *a*, the area of wall associated with the tie structure, is $2(4\sqrt{2}ax\sqrt{2})\gamma$, so that the wall energy is $16ax\gamma$.

A 180° wall total area *A*, moving through a distance *x* reverses

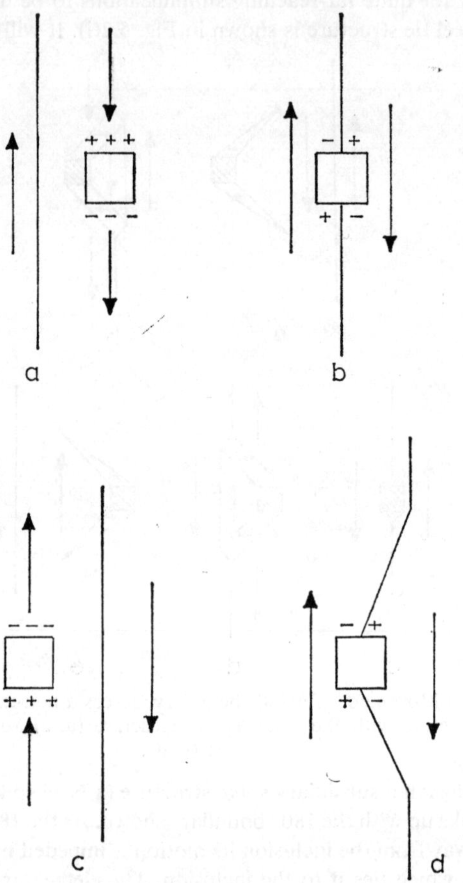

FIG. 5.2 (ii) a, b, c Movement of a 180° boundary across an inclusion with free-pole formation.

the magnetization over a volume Ax, and the magnetic energy term is $2HI_sxA$, so that

$$E' = E - HI = \frac{16a\gamma x - 2HI_sxA}{V} \tag{5.2a}$$

which is in equilibrium when

$$H = \frac{16a\gamma}{2I_sA} = \frac{8a\gamma}{AI_s} \tag{5.2b}$$

i.e. the energy curve (E, x) is linear. In fact this expression holds only for large values of x; for small values of x the curve is as shown. This, however, indicates that the condition of eqn (5.1b) would be satisfied as soon as x becomes large and that the 180° boundary will then break away, i.e.

$$H_c = \frac{8\gamma a}{AI_s} \tag{5.2c}$$

Writing $\alpha = a^2/A$ as the fractional area of the boundary taken up by the inclusion,

$$H_c = \frac{8\alpha\gamma}{I_sa} \tag{5.2d}$$

with $\gamma = 1$ erg cm^{-2}, $I_s = 10^3$ e.m.u. cm^{-3}, $\alpha = 0{\cdot}1$, then for H_c approximately 1 oersted, a is approximately 10^{-4}cm.

The alternative configuration is that shown in Fig. 5.2(ii)b with a 180° boundary intersecting a cubic non-magnetic inclusion without the formation of a subsidiary domain structure. The magnetostatic energy associated with this configuration is approximately one half that of Fig. 5.2(ii)a. The energy associated with Fig. 5.2(ii)a is equal to the demagnetizing energy of a uniformly magnetized cube

$$\tfrac{1}{2}(\tfrac{4}{3}\pi I_s^2a^3)$$

so that the energy expression for Fig. 5.2(ii)b is

$$\tfrac{1}{3}\pi I_s^2a^3 \tag{5.3a}$$

The curve for the variation in energy of the boundary as it moves across the inclusion will in fact be similar to Fig. 5.1b, but it is

sufficient for this discussion to assume that as the boundary moves away from the centre of the inclusion the energy is proportional to x, i.e.

$$\left(\frac{2}{3}\pi I_s^2 a^3\right)\left(\frac{x}{a}\right). \tag{5.3b}$$

This term replaces the term for the wall energy in the previous discussion, eqn (5.2a), giving an expression for the coercivity

$$H_c = \tfrac{1}{3}\pi I_s \alpha. \tag{5.3c}$$

This expression gives a value of the coercivity which is much higher than that given by eqn (5.2d) by a factor for iron of $4 \times 10^5 a$. Only for inclusions whose dimension a is of the order 10^{-6} cm are the terms comparable; the geometry of this model can be modified to produce a more favourable situation energetically such as that in Fig. 5.2(ii)d where the 180° boundary is distorted near the inclusion; the result is however still essentially the same, that the energy is high and that this is reduced by the formation of the Néel tie structure for larger inclusions, i.e. a greater than 10^{-4} cm. The general result of a number of analyses of this problem is that inclusions whose dimensions are approximately equal to the thickness of the domain wall are most effective in producing hysteresis. The significance of this is that the inclusions are so small that the Néel tie structure cannot be formed so there is no means of reducing the magnetostatic energy and the coercivity is therefore high. Unfortunately experimental observation on this scale, even with an electron microscope, is difficult and there is no direct confirmation of this.

The effect of internal stresses in introducing an additional anisotropy must also be taken into account in a consideration of hysteresis and coercivity. This additional anisotropy in the simplest example $K + K_\sigma = K + \frac{3}{2}\lambda_s\sigma$ will obviously affect the boundary energy. Equally important are variations in stress which give rise to fluctuations in the direction of magnetization, producing additional localized fields. Néel has carried out a most sophisticated analysis of this effect in which he allows for the presence of inclusions which in an analogous fashion result in fluctuation in the magnitude (rather than the direction) of the magnetization, so giving rise to the free-pole effects discussed above. The expressions for the coercivity obtained by Néel are for iron

$$H_c = 2{\cdot}1v_\sigma + 360v_i \tag{5.4a}$$

and for nickel

$$H_c = 330v_\sigma + 97v_i \tag{5.4b}$$

where v_i and v_σ are the effective fractional volumes of the specimen occupied by inclusions and regions of stress. Clearly stress is a predominant factor in nickel and inclusions in iron; this results from the much higher magnetostriction coefficient of nickel.

These values of H_c derived from the above expression are not unreasonable, but it is obvious that even if the expressions for H_c are acceptable much information concerning the size and distribution of inclusions is required; to make any really significant estimate of H_c it would be necessary to carry out a summation for all inclusions and boundaries. In practice there is so much latitude in the choice of parameters in most expressions for the coercivity that assessment of their significance is virtually impossible. The main virtue of the development of the formulae for the coercivity is in the insight gained from the analysis of the magnetization process.

5.2 The Néel block

Much of what has been said about domain structure concerns the behaviour of the material in low fields, i.e. over that part of the hysteresis curve near the coercive field. The general picture of the magnetization curve is that in the steep part of the curve the magnetization changes by means of 180° boundaries moving across the specimen. At higher fields, above the knee of the curve, much of the domain structure and domain boundaries has disappeared and changes in magnetization take place mainly by rotation. The process in high fields has been examined in more detail by Néel who has considered the domain structure in an iron single crystal as shown in Fig. 5.3. This domain structure consists of a series of slabs in which the magnetization lies parallel to one or other of the easy ⟨100⟩ directions, with the domains separated by 90° boundaries (Fig. 5.3c). At the edge of the block the structure is completed by

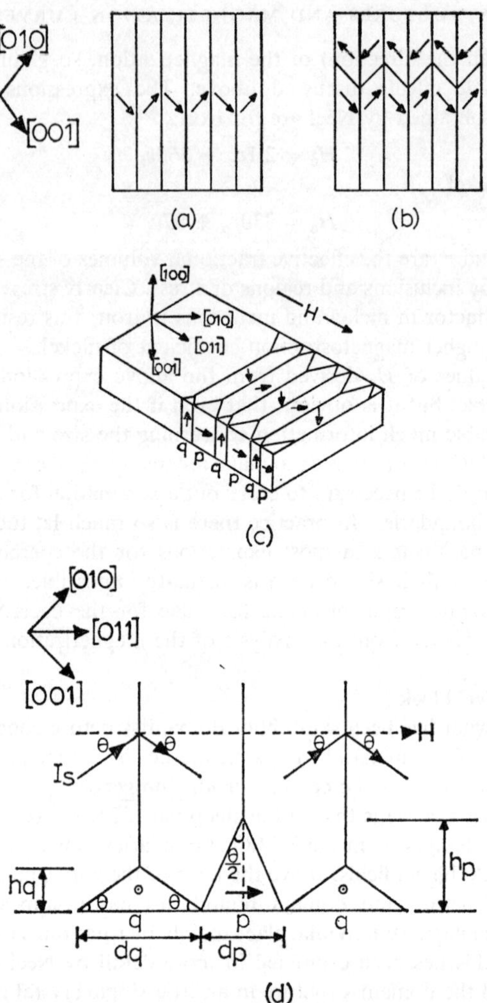

FIG. 5.3 (a–d) Domain structure in a Néel block. (a) and (b) Reversal of magnetization by the movement of 180° domain wall across the 90° domain structure. (c) Single crystal orientation for the Néel block showing the closure domain structure p and q. (d) Details of the structure of the p and q domains.

'closure' domains (Fig. 5.3d). (This configuration does not completely 'close' the magnetic flux, and in fact other structures have been observed experimentally. At low fields the magnetization changes by the reversal of the magnetic vectors associated with the movement of a 180° boundary across the specimen, as shown in Figs. 5.3a and b. The domain structure consisting of 90° boundaries is retained in higher fields, and the magnetization then changes by rotation of the magnetization vector out of the easy direction and by the associated movements of the 90° boundaries of Fig. 5.3c. From the changes in the domain spacing the changes in magnetization at high fields can be estimated. An example of this type of structure is shown in plate O.

The crystal is in the form of an infinitely long strip width L, and thickness t, with the flat surface lying in the (100) plane, and with its long axis parallel to the [011] axis. There are thus two easy ⟨100⟩ directions in the plane of the strip. The domain structure in a zero-applied field takes the form of layers, spacing $\frac{1}{2}$d, lying parallel to the (011) plane, magnetized as shown in the [010] and [001] directions, to form 90° domains so that no free pole arises on the interdomain surfaces.

With the application of a magnetic field in the [011] direction the magnetization in these domains turns out of the easy directions, rotating against the magnetocrystalline anisotropy but still with the normal component of the magnetization continuous across the boundary. The energy expression due to this rotation is (eqn 3.14)

$$E' = E - \mathbf{H}.\mathbf{I} = K_1 \cos^2\left(\frac{\pi}{4} + \theta\right) \cos^2\left(\frac{\pi}{4} - \theta\right) - HI_s \cos\theta$$

$$= \tfrac{1}{4}K_1 \cos^2 2\theta - HI_s \cos\theta \tag{5.5a}$$

where θ is the angle between I_s and the [011] direction. This energy is a minimum when (writing $K_1 = K$)

$$HI_s = 2K \cos\theta \cos 2\theta \tag{5.5b}$$

giving the minimum energy

$$E'_{\min} = -\tfrac{1}{4}K \cos 2\theta(6\cos^2\theta + 1) \tag{5.5c}$$

θ being the equilibrium value for a specified value of H. Free-pole effects at the edges of the crystal are reduced by the formation of subsidiary prismatic structures p and q; Néel shows that the energy of these closure domains is a minimum when the magnetic vectors lie approximately parallel (p) and at right angles (q) to the magnetic field as shown in Fig. 5.3d. The domain walls are oriented so that the normal component of the magnetization is continuous across a boundary; i.e. the walls of the q-domains lie parallel to magnetization vectors in the main domains; the normals to the p-walls make an angle $\frac{1}{2}\theta$ with the main magnetization vectors. The energy per unit volume of the p-domains is (cf. eqn (5.5a))

$$K \sin^4 \frac{\pi}{4} - HI_s = \tfrac{1}{4}K - 2K \cos\theta \cos 2\theta \tag{5.6}$$

from eqn (5.5b); the corresponding terms for the q-domains are zero. The difference between the energy density of the prismatic domains and that of the main domains (eqn (5.6)), i.e. the energy of formation is (per unit volume)

$$\begin{aligned} E_p' &= \tfrac{1}{4}K - 2K\cos\theta\cos 2\theta + (\tfrac{1}{4}K\cos 2\theta)(6\cos^2\theta + 1) \\ &= (K\cos\theta)(1 - \cos\theta)^2(2 + 3\cos\theta) \end{aligned} \tag{5.7}$$

and $$E_q' = (\tfrac{1}{4}K\cos 2\theta)(6\cos^2\theta + 1). \tag{5.8}$$

The dimensions of the closure domains are

$$h_p = \tfrac{1}{2}d_p \cot\frac{\theta}{2} = (\tfrac{1}{2}d_p \sin\theta)(1 - \cos\theta)^{-1} \tag{5.9}$$

and $$h_q = \tfrac{1}{2}d_q \tan\theta \tag{5.10}$$

with the volumes of $\frac{1}{2}d_p h_p t$ and $\frac{1}{2}d_q h_q t$. The energy of formation of a single prism closure domain is when

$$w_p = \tfrac{1}{4}d_p^2 Kt \sin\theta\cos\theta\,(1 - \cos\theta)(2 + 3\cos\theta) = d_p^2 t K_p \tag{5.11a}$$

$$w_q = \tfrac{1}{16}d_q^2 Kt \tan\theta\cos 2\theta\,(6\cos^2\theta + 1) = d_q^2 t K_q \tag{5.11b}$$

where K_p and K_q are anisotropy factors for the p and q configurations. The energy of a pair of closure domains is then

$$w_p + w_q = K_p d_p^2 t + K_q d_q^2 t \tag{5.12}$$

which is a minimum for a given $D = d_p + d_q$ when

$$d_p = \frac{K_q D}{K_p + K_q} \quad \text{and} \quad d_q = \frac{K_p D}{K_p + K_q}. \tag{5.13a}$$

Substituting for eqn (5.13a) in eqn (5.12) gives

$$w_p + w_q = tD^2K_pK_q(K_p + K_q)^{-1}. \tag{5.13b}$$

The total energy for the Néel block is then per unit area of section (tD)

$$\frac{2\gamma L}{D} + 2DK_pK_q(K_p + K_q)^{-1}, L \gg D, \tag{5.14}$$

which is a minimum when

$$D^2 = (d_p + d_q)^2 = \frac{\gamma L(K_p + K_q)}{K_pK_q} \tag{5.15a}$$

$$= \frac{\gamma LD}{d_pK_p} = \frac{\gamma LD}{d_qK_q} \tag{5.15b}$$

From 5.15a the domain spacing is, as in simpler configurations, proportional to $L^{\frac{1}{2}}$; although this has been confirmed experimentally there is a discrepancy in the constant of proportionality. This is probably due to the formation of a more complicated structure than the p and q domains, and this would reduce the energy of the system.

The volumes of the p and q domains are

$$V_p = \tfrac{1}{2}d_ph_pt = \tfrac{1}{4}d_p{}^2t \sin\theta(1 - \cos\theta)^{-1}$$
$$= \frac{\gamma^2L^2t}{4K_p{}^2D^2}\frac{\sin\theta}{(1 - \cos\theta)} \tag{5.16}$$

$$V_q = \tfrac{1}{2}d_qh_qt = \tfrac{1}{4}d_q{}^2t\tan\theta = \frac{\gamma^2L^2t\tan\theta}{4K_q{}^2D^2}. \tag{5.17}$$

The resultant magnetization is then, taking the component of the magnetization parallel to H,

$$\frac{I}{I_s} = \frac{V\cos\theta + V_p}{V + V_p + V_q}, \tag{5.18a}$$

where V is the volume of the primary domains in one half of the specimen associated with one p domain and an adjacent q domain so that $V + V_p + V_q = \tfrac{1}{2}LDt$ (the other half of the primary domain is taken for the p and q domains on the opposite side). There is no term in V_q in the numerator of eqn (5.18a) as the magnetization I_q is normal to the field; the expression for V_q is still required, as a change in the volume V_q implies a change in V_p i.e. $V. I/I_s$ is evaluated from eqn (5.18a) in the form.

$$\frac{I}{I_s} = \frac{(\frac{1}{2}LDt - V_p - V_q)\cos\theta + V_p}{\frac{1}{2}LDt}$$

$$= \frac{\frac{1}{2}LDt\cos\theta - V_q\cos\theta + V_p(1 - \cos\theta)}{\frac{1}{2}LDt} \quad (5.18b)$$

and substituting from eqns (5.16), (5.17) and for K_p, K_q from eqns (5.13a) and (5.15),

$$\frac{I}{I_s} = \cos\theta + \frac{\gamma(K_q - K_p)\sin\theta}{2DK_pK_q}. \quad (5.18c)$$

In order to determine the magnetization curve we proceed as follows. For an applied field the equilibrium value of θ is determined graphically or by interpolation from eqn (5.5b). i.e.

$$\cos\theta\cos 2\theta = \frac{HI_s}{2K};$$

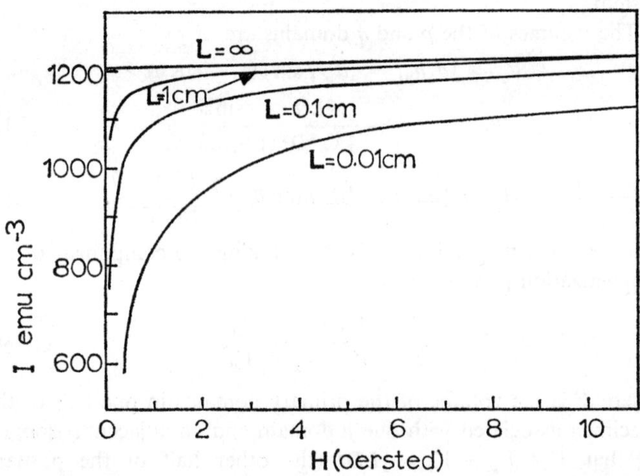

FIG. 5.4 Magnetization curves calculated for a Néel block width L, thickness $t = 1$ cm.

thus K_p and K_q are derived from eqns (5.11a) and (5.11b) and D follows from eqn (5.15). Hence I/I_s as a function of H can be constructed for various values of L. Examples of magnetization curves determined in this way are shown in Fig. 5.4.

The difference between the measured (I, H) curves and those calculated in the above way is large and is probably due to the nature of more complicated closure domains which take the place of the simpler structures.

5.3 The nucleation of magnetic domains

Although the formation of a particular domain configuration in a specimen represents a state of low magnetic energy the process of domain formation may involve processes with relatively high energies. Whatever the mechanism by which the domains are formed it would require some means of producing a domain wall and this involves the rotation of the magnetization vector against the exchange and magnetocrystalline forces. The exchange field for iron is equivalent to a magnetic field of order 10^7 oersted, and localized fields approaching this magnitude would be required to nucleate a domain wall. It is to be expected that large magnetostatic fields must be produced near cracks and flaws at the surface of the specimen or at grain boundaries; these might be of order $4\pi I$, i.e. 2×10^4 oersted iron, or possibly higher but clearly too small for domain nucleation in the terms described.

Clear and definite information is not easy to obtain because of the difficulty of experimental observation on a process which takes place on the scale of a domain wall thickness (10^2 to 10^3 Å). In a typical investigation the specimen is magnetically saturated in a high field, and all visible signs of a domain structure eliminated. The field is then reduced to zero and reversed until the domain structure is again formed; it is usually found that when the experiment is repeated the domains nucleate at the same sites (although of course the nucleation process itself cannot be observed). This must be because (i) the localized magnetic field at the nucleation site is large enough to produce nucleation or (ii) the domain nucleii have not been eliminated even in high fields. Alternative (ii) arises naturally if it is assumed that domains cannot be nucleated in fields available in the laboratory, as it then follows that they cannot be destroyed.

In thin films domains are seen to nucleate particularly easily at the edge of the film; in addition, domain boundaries are sometimes observed to form as a development of ripple structure which is due to local deviations in the anisotropy.

Although domain nucleation by thermal agitation of the electron spins is unlikely in most materials at room temperature, it must be borne in mind that at some stage in their preparation many ferromagnetic specimens are heated or sintered at high temperatures and in some cases cooled through the Curie point. Nucleation could take place at high temperatures, and the domain nucleii would then be retained at room temperature with little chance of being destroyed in the normal process of magnetization. Other specimens are subject to severe mechanical working and straining which could provide a means of nucleation. In specimens such as thin films prepared by evaporation or chemical deposition the process of growth from small particles might allow of rotation of the spin vectors by thermal agitation.

It will be seen from the number of possibilities outlined here that there is much to be learned about the nucleation of magnetic domains; it is also obvious that there is, of course, no reason why there should be only one single mechanism involved in the formation process in all the types of specimen.

6: Single Domain Particles

In a sense it could be claimed that this chapter has no place in a book on domains in that it deals with those specimens which, because of their size and shape, are too small for the formation of a domain structure. The reason for this is that a relatively large part of the specimen would be taken up by a domain wall of the usual type which is essentially a region of magnetic stress or high energy. In a larger specimen the volume taken up by a domain boundary is relatively small and the contribution to the energy is small.

The phenomenon of the single domain particle is important because in the absence of a domain structure the coercivity must be high. The essential point about domain formation is that it is a means of reducing the energy of the system and hence of reducing the coercivity. Where a high coercivity material is required, as for instance for a permanent magnet, domain formation should be avoided. The problem of a single domain particle is thus one of some commercial importance.

6.1 Particle size

(i) *High anisotropy*

In the estimation of the size of single domain particles it is convenient to consider separately high anisotropy materials with narrow boundaries and low anisotropy materials with wide boundaries. To derive the condition for the single domain formation the elementary example of a cube, side D, is first considered. Such a specimen, if it consisted of a single domain could be magnetized as shown in Fig. 6.1c with the demagnetizing energy for the volume D^3

$$\tfrac{1}{2}NI^2D^3 = \tfrac{2}{3}\pi I^2D^3 \tag{6.1a}$$

assuming the cube to be uniformly magnetized so that its demagnetizing factor is equal to that of a sphere. A possible domain structure for a material with cubic anisotropy is shown in Fig. 6.1d;

others might be considered but the difference would not be significant. The boundary width is considered to be small compared to the dimensions of the specimen. The energy of the structure shown will be that of the boundary walls $2D^2\sqrt{2}\gamma$.

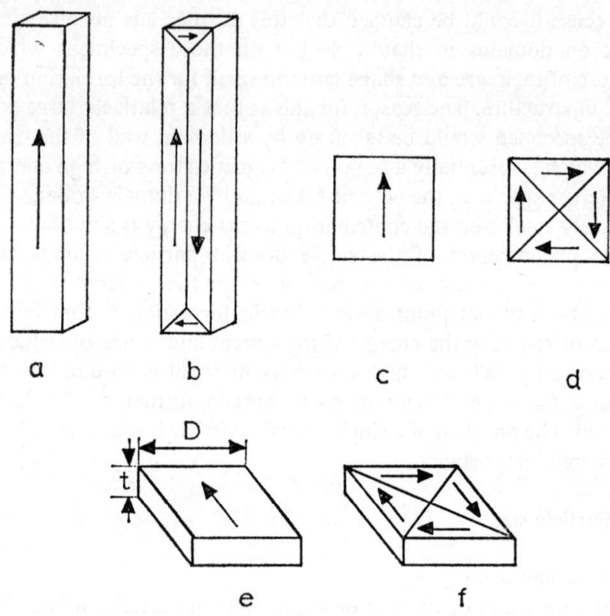

FIG. 6.1 (a–f) Alternative configurations for 'single domain' particles.

This energy is higher than that of the uniformly magnetized single domain particle when

$$2D^2\sqrt{2}\gamma > \tfrac{2}{3}\pi I_s{}^2 D^3 \tag{6.1b}$$

giving a critical size for the particle

$$D = \frac{3\sqrt{2}\gamma}{\pi I_s{}^2}. \tag{6.2}$$

Thus for a cubic specimen there is a minimum size given by eqn (6.2) below which the formation of the domain structure of the type illustrated in Fig. 6.1d is energetically unfavourable; for $\gamma = 1$ erg cm^{-2} and $I = 1700$ e.m.u. cm^{-3} (iron), $D = 45$ Å.

For long needle-like specimens the alternative configurations are shown in Figs. 6.1a and b. The demagnetizing energy term is $\frac{1}{2}NI_s^2D^2L$, and the domain wall term γLD, gives for the critical diameter

$$D = \frac{2\gamma}{NI_s^2} \tag{6.3}$$

Clearly the critical diameter increases with decreasing demagnetizing factor, i.e. with increasing length to diameter ratio. For an iron needle, length to diameter ratio $q = 10$, $D = 250$ Å and for $q = 20$, $D = 800$ Å.

The model can only be applied where the material has cubic anisotropy and with a high value of K so that the boundary thickness is small, much less than the thickness of the specimen. Although the value of K for iron is not high enough to be justified, these calculations are useful in finding orders of magnitude.

If a uniaxial material is considered, the energy for the alternative structures is given by (Fig. 6.1c)

$$\tfrac{2}{3}\pi I_s^2D^3 \quad \text{and} \quad \tfrac{1}{2}.\tfrac{2}{3}\pi I_s^2D^3 + D^2\gamma \tag{6.4a}$$

assuming the demagnetizing factor is halved by subdivision into two antiparallel domains. The critical value of D is

$$D = \frac{3\gamma}{\pi I_s^2} \tag{6.4b}$$

not significantly different from eqn (6.2b).

For cobalt this gives, with $\gamma = 6$ erg cm^{-2} and $I_s = 1400$, $D = 300$ Å compared with the value for the boundary thickness $\delta = 150$ Å. For the long needle of Fig. 6.1a the demagnetizing energy is as before $\frac{1}{2}NI_s^2D^2L$ and the energy of the two domain structure (without the closure domains) is $DL\gamma + \frac{1}{2}(\frac{1}{2}NI_s^2D^2L)$ giving for the critical value for D

$$D = \frac{4\gamma}{NI_s^2}. \tag{6.5}$$

For a thin film in the shape of a square magnetized in the plane of the film the corresponding equation is Figs. 6.1e and f

$$2\sqrt{2}Dt\gamma = \tfrac{1}{2}NI_s^2tD^2 \qquad (6.6)$$

with $N = \pi^2 t/D$ (the expression for the oblate spheroid, eqn 3.4d); this gives

$$t = \frac{4\sqrt{2}\gamma}{\pi^2 I_s^2}. \qquad (6.7)$$

For thin films in the range of thickness being considered here, γ is relatively high, $\simeq 5$ erg cm^{-2}, because of the magnetostatic energy contribution (see Chapter 7); with for Permalloy $I_s = 800$ e.m.u. cm^{-3}, $t \approx 500$ Å. For a uniaxial material the energy of the two-domain structure is (cf. eqn 6.4a)

$$\tfrac{1}{2}.\tfrac{1}{2}NI_s^2tD^2 + Dt\gamma$$

with $N = \pi^2 t/D$ giving

$$t = \frac{4\gamma}{\pi^2 I_s^2}. \qquad (6.8)$$

(ii) Low anisotropy

Where the anisotropy is low the width of a domain wall will be relatively large and the critical condition can be considered to be when the wall width δ takes up the whole width D of the specimen.

If the exchange energy contribution is equal to that of the anisotropy the energy density is then (cf. eqn (4.3) with the two terms equal) $2A(\mathrm{d}\phi/\mathrm{d}x)^2 \approx 2A(\pi/D)^2$. It is assumed that the magnetization vectors in the boundary rotate uniformly through the specimen width D, i.e. following the asymptotic line of Fig. 4.1b. The demagnetizing energy density is $\tfrac{1}{2}NI_s^2$ so that the critical diameter is given by

$$\tfrac{1}{2}NI_s^2 = 2A\left(\frac{\pi}{D}\right)^2 \qquad (6.9)$$

$$D = \frac{2\pi}{I_s}\left(\frac{A}{N}\right)^{\frac{1}{2}} \qquad (6.10)$$

where N is the demagnetizing factor in the direction parallel to the domain boundary. For a length to diameter ratio $q = 20$, $D = 1800$ Å and for $q = 10$, $D = 1000$ Å for iron.

6.2 Coercivity of single domain particles

A spherical isotropic single domain particle ($K = 0$) can be magnetized with equal ease in any direction and the magnetization vector can be rotated from one direction to another by the application of an infinitesimally small field so that the coercivity is zero.

To produce a high coercivity, or any coercivity for that matter, some anisotropy must be present, so that there is a preferred direction of magnetization: a finite field is then required to turn the magnetization away from this easy axis. This anisotropy may take the form of (ia) magnetocrystalline anisotropy, (ib) strain anisotropy, (iii) shape anisotropy with the specimen longer in one direction so that the demagnetizing fields in that direction are less than in another. It is necessary only to consider in detail (ia) and (iii) as (ib) can be allowed for as a modification of (ia); clearly there could also be a combination of any of these.

(i) *Magnetocrystalline anisotropy*

Consider the spherical specimen of Fig. 6.2a without domain structure and with only positive uniaxial magnetocrystalline anisotropy. The variable part of the magnetic energy is $HI_s \cos \phi$ and that due to anisotropy is $K \sin^2 \phi$. The total energy is then

$$E = E_K - \mathbf{H}.\mathbf{I} = K \sin^2 \phi - HI_s \cos \phi \qquad (6.11a)$$

giving the equilibrium condition when $dE/d\phi = 0$, i.e. when

$$(2K \cos \phi + HI_s) \sin \phi = 0, \qquad (6.11b)$$

with the condition for stability $d^2E/d\phi^2 > 0$.

i.e.
$$2K \cos 2\phi + HI_s \cos \phi > 0. \qquad (6.11c)$$

Consider a cycle commencing with H large and positive, reducing to zero then reversing and increasing to a large negative value of H and back. With H positive the system is stable with $\phi = 0$ for all values of H, i.e. with H and I parallel. On reversing H, i.e. H negative, the system remains stable, with $\phi = 0$, until $H = -(2K/I_s)$ when it becomes unstable and the magnetization reverses to give $\phi = \pi$. The system is again stable (with eqn (6.11c) satisfied) and remains so as the negative field is further increased. On continuing the cycle a 'square' hysteresis loop is described.

The value of $H = 2K/I_s$ is the maximum coercivity H_K, due to magnetocrystalline anisotropy, of a specimen originally magnetized parallel to an easy direction; i.e. this is the negative field required to reverse the magnetization; where the anisotropy is due to magnetostrictive strain, K is replaced by $\frac{3}{2}\lambda_s\sigma$ and H_K by $H_\sigma = 3\lambda_s\sigma/I_s$.

The more general case is shown in Fig. 6.2b with the applied field

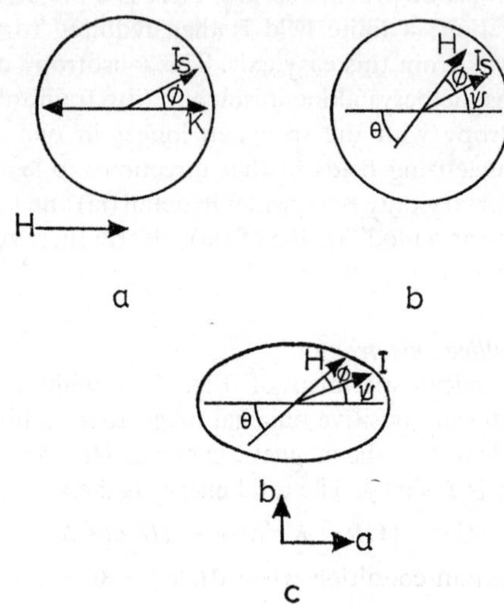

FIG. 6.2 (a–c) Orientation of magnetization and field in single domain particles: (a) and (b) spheres with magnetocrystalline anisotropy, (c) ellipsoid with shape anisotropy only.

H in a direction at an angle $\theta = \phi + \psi$ with the easy direction K. The expression for the energy is then

$$\begin{aligned} E &= E_K - \mathbf{H}.\mathbf{I} \\ &= K \sin^2 \psi - HI_s \cos \theta \qquad (6.12) \\ &= K \sin^2 (\phi - \theta) - HI_s \cos \phi. \end{aligned}$$

The conditions for equilibrium are given by

$$\frac{dE}{d\phi} = K \sin 2(\phi - \theta) + HI_s \sin \phi = 0 \qquad (6.13a)$$

corresponding to

$$\frac{d_2E}{d\phi^2} = 2K\cos 2(\phi - \theta) + HI_s\cos\phi \tag{6.13b}$$

positive for minima and negative for maxima.

The magnetization curves for various orientations of H, I_s and K have been derived numerically from eqn (6.13). The magnetization measured in the direction of the field is $I = I_s \cos\phi$ and as I_s is a constant for a given material, $\cos\phi$ is a measure of the 'reduced' magnetization I/I_s. This is shown in Fig. 6.3a (plotted against the reduced field $h = (H/H_K)$.

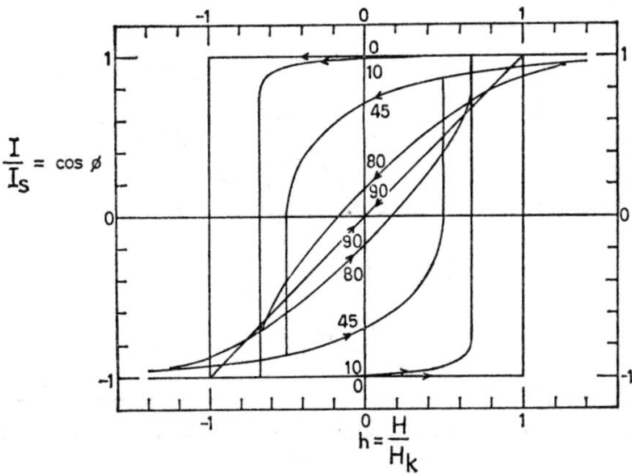

FIG. 6.3a Magnetization curves for spherical single domain particles with uniaxial anisotropy K with $H_K = 2K/I_s$. The curves also apply to prolate spheroids ($K = 0$); $h = H/H_s$, $H_s = (N_b - N_a) I_s$, where N_b and N_a are the demagnetizing factors along the polar and equatorial axes. The angle θ between the easy direction and the field direction is shown in degrees on the curves. (Stoner and Wohlfarth, 1948.)

It will be seen that the hysteresis curves range from rectangular loops to straight lines showing no hysteresis, depending on the particle orientation. The rectangular loop corresponds to the conditions discussed in equations (6.11).

For an assembly of particles randomly oriented, and without interaction between particles, the hysteresis loop is as shown in Fig. 63b. The coercivity H_c, initial susceptibility κ_0 and remanence I are

$$H_c = 0{\cdot}479 H_K \tag{6.14a}$$

$$\kappa_0 = \frac{2}{3}\left(\frac{I_s}{H_K}\right) \tag{6.14b}$$

$$I_r = \tfrac{1}{2} I_s \tag{6.14c}$$

(ii) *Shape anisotropy*

To produce a change in magnetization in the ellipsoidal single-domain particle of Fig. 6.2c from a direction $+x$ to $-x$ the change

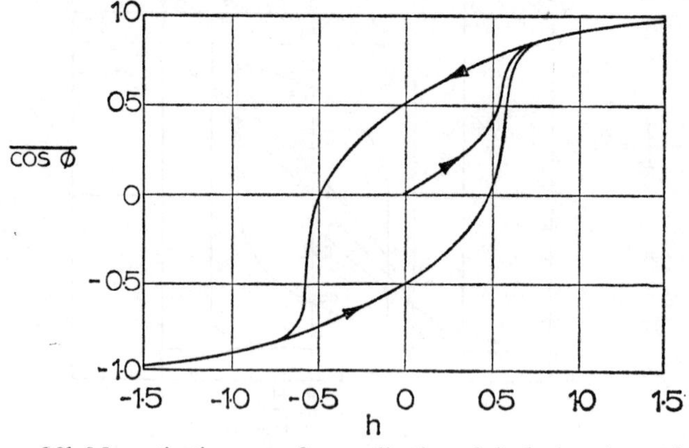

FIG. 6.3b Magnetization curve for a collection of single domain particles (see Fig. 6.3a) oriented at random. $\cos\phi = I_m/I_s$ where I_m is the mean magnetization resolved parallel to H. (Stoner and Wohlfarth, 1948.)

in applied field must be sufficiently great to turn the magnetic vector from the direction (x) of minimum demagnetizing factor $N = N_a$ through the direction of maximum $N = N_b$. The change in field is thus $(N_b - N_a)I_s$ and this is the maximum coercivity H_c of a single domain particle in the form of a prolate ellipsoid of revolution ($b = c < a$).

The demagnetizing factor when the particle is magnetized as

shown is

$$N_a \cos^2 \psi + N_b \sin^2 \psi$$

and the energy

$$\tfrac{1}{2}I_s^2(N_a \cos^2 \psi + N_b \sin^2 \psi). \qquad (6.15a)$$

Clearly, this can be put into the same form as the expression for the magnetocrystalline energy $K \sin^2 \psi$,

i.e.
$$\tfrac{1}{2}I_s^2(\overline{N_a + N_b - N_a \sin^2 \psi});$$

using the variable part of the expression (6.15b) this gives in place of H_K,

$$H_s = (N_b - N_a)I_s \qquad (6.16)$$

with hysteresis curves as in Fig. 6.3a and $h = H/H_s$.

The corresponding expressions for coercivity, susceptibility and remanence in place of eqn (6.14) are

$$H_c = 0{\cdot}479(N_b - N_a)I_s \qquad (6.17a)$$

$$\kappa_0 = \frac{2}{3}\left(\frac{1}{N_b - N_a}\right) \qquad (6.17b)$$

$$I_r = \tfrac{1}{2}I_s \qquad (6.17c)$$

and the hysteresis loop for this assembly is shown in Fig. 6.3b. For an oblate ellipsoid of revolution ($b = c > a$) the position is complicated by the fact that the magnetization vector can rotate in the equatorial plane in which N is constant, so that no energy is required; the coercivity is then zero. This is of some importance as there is evidence that in some high-coercivity materials the precipitate particles are approximately disc-shaped; they are thus expected to have zero coercive force. One possible explanation may be that the discs may have non-circular sections when some degree of shape anisotropy would again be important.

6.3 Permanent magnet materials

A study of modern permanent magnetic materials is intimately related to the subject of single-domain particles. The early permanent magnets were made of hardened carbon steels and cobalt steels which were hard both physically and magnetically. These alloys, which were quenched during their preparation, had a crystalline structure consisting of regions of the high temperature γ (fcc) phase, mixed with the α (bcc) low temperature phase. As a result the material was highly strained internally and this, together with the inclusions present, impeded the domain wall motion and the process of magnetization reversal to give a relatively high coercivity of the order 100–200 oersteds.

The assessment of the utility of permanent magnetic materials requires rather more than the coercivity; the significant part of the hysteresis curve is the quadrant B to H in Fig. 6.4 as it is on this part of the curve that the permanent magnet is operated. The effect

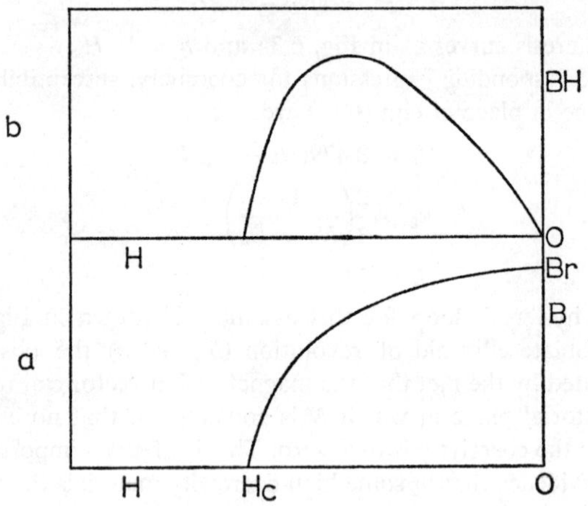

FIG. 6.4 (a–b) (a) Part of a magnetization curve between H_c and B of a permanent magnet material – 'the demagnetization curve'; (b) above is a curve of (BH) against H showing a maximum as $(BH)_{max}$.

of demagnetizing fields is to move the specimen from the remanent point B_r along the (B, H) curve until at a large enough reverse field $H = H_c$, the specimen is demagnetized. Not only must H_c be large but B_r must be also large to give a high value of the attractive force or lifting power. The parameter used as a measure of utility is the product BH, i.e. $(BH)_{max}$; a value for the cobalt steels mentioned above would be $\sim 10^6$ gauss oersted. In the production of really high values of $(BH)_{max}$ it is necessary to avoid the formation of domain walls so that the only means of reversing the magnetization is by rotation against anisotropic forces which are designed to be as large as possible. Permanent magnet materials fall into two groups; those whose properties are due mainly to shape anisotropy and those whose properties are due to crystalline anisotropy.

Shape anisotropy

The permanent magnet alloys of the type known as Alnico, Alcomax or Ticonal are alloys with a typical composition 14% nickel, 8% aluminium, 20% cobalt, 3% copper and the balance of iron, but there are variations in this in different applications.

These alloys are prepared by quenching from about 1200°C then subsequently annealing. As a result of the annealing a two-phase system is produced, one phase strongly magnetic and the other non-magnetic. The structure consists essentially of magnetic single-domain particles, some 10^3 Å long, in a non-magnetic matrix. The particles are elongated or plate-like, and this shape anisotropy gives rise to high coercivities of 500 oersted, remanence 5000 to 8000 gauss and $(BH)_{max}$ $1{\cdot}5 \times 10^6$. By cooling or annealing in a magnetic field the particles are aligned more or less with the long axis parallel to the field so that the material is anisotropic, increasing the remanence to some 13,000 gauss and $(BH)_{max}$ to 8×10^6; the coercivity is not greatly increased. This improvement in permanent magnet properties is in the direction of the applied field only; there is a reduction in the parameters in the direction at right angles to the field.

It should be emphasized that although the particles are orientated in these anisotropic alloys the crystal orientation of the particles is random. Further improvement is obtained if crystal growth in the direction of the applied field can be induced by extracting heat in one direction during the growth process. These 'columnar' magnet materials have even higher coercivities as a result of the addition of an easy crystallographic direction of magnetization to that due to the shape of the specimen. The effect of crystalline anisotropy is not as high as might be expected with a material containing cobalt, which in the normal room-temperature phase is hexagonal with a high value of K. In the form of very small particles any cobalt would be in the fcc form which has cubic anisotropy with a much lower value of K.

Many attempts have been made to produce permanent magnet materials by agglomerating elongated single-domain particles of iron or cobalt. Although some of these have been successful there are many manufacturing difficulties and the expense is so great that they have only limited applications.

Crystalline anisotropy

The materials of importance under this heading are a group of mixed oxides known as hexagonal ferrites of which barium ferrite, $BaFe_{12}O_{19}$, is the best known; it is also marketed under the name of Ferroxdure or Hexagonal M compound. The crystalline structure is hexagonal having a highly preferred axis of magnetization along the c (hexagonal) axis with $K = 3{\cdot}3 \times 10^6$ erg cm^{-3}; since $I_s = 380$ e.m.u. cm^{-3} this gives a relatively high value for $H_K = 17{,}000$ oersted.

The magnet material is prepared in finely divided powder form by milling and it is then compressed into shape and sintered giving $B_r = 2000$ gauss, $H_c = 1700$ oersted, $(BH)_{max} = 1{\cdot}0 \times 10^6$. The particles may be aligned in a strong magnetic field during compression to give an anisotropic product when the value of the remanence is doubled and $(BH)_{max}$ trebled in the field direction.

Incoherent rotation

The problem so far has to some extent been oversimplified by classifying materials under two headings of shape and crystalline anisotropy. It is obvious that both effects would be present to some extent in any of these materials. In addition, the whole mechanism is certainly modified by interaction between the particles. An example of this interaction is the 'fanning' effect illustrated in Fig. 6.5a which would produce an alternative means of reversing and of rotating the magnetization and result in a lower coercivity.

This is an example of incoherent rotation in which the spins no longer remain parallel during rotating or reversal; it has been assumed so far that the spins rotate coherently. In the individual particles themselves there are other methods of changing the magnetization, for example 'buckling' and 'curling' which involve incoherent rotation. In buckling (Fig. 6.5b) the magnetization rotates with a sinusoidal variation in the rotation producing a strain effect rather similar to a strut under compression; in curling the magnetization rotates so as to give a circumferential component.* Although both of these processes increase the exchange contribution to the energy they can result in a reduction of coercivity compared with the coherent rotation model.

A very readable account of permanent magnet materials is given in *Attraction and Repulsion* by M. McCaig (1967).

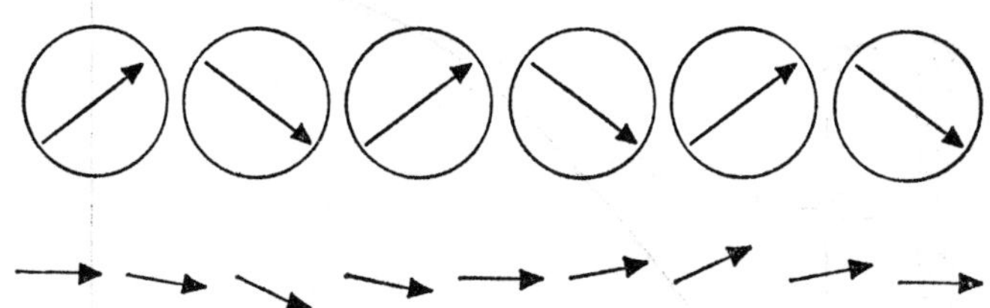

FIG. 6.5 (a–b) Examples of incoherent rotation of the magnetization vectors in single domain particles: (a) 'fanning', (b) buckling.

6.4 Magnetic recording tapes

These must have a high coercivity of 100's of oersteds. The tape is then insensitive to the effects of stray fields and can be effectively magnetized only by a strong magnetic field such as can be produced by a recording head. The recording head is in effect a small electromagnet with a ferrite core which permits the production of rapidly changing localized magnetic fields over a small area of the tape.

The materials suitable for use as films on recording tapes are the permanent magnet materials with single-domain particle structure. There would therefore appear to be a wide choice of alloys and compounds but most tapes are made with γFe_2O_3, maghemite, in a plastic

* This is difficult to illustrate, but the circumferential component in the basal plane would be as in Fig. 7.8a. In a long cylinder, for instance, the main magnetization may lie in the long direction, but there is a circumferential component, possibly like side spin on a projectile. The magnetization rotates incoherently as the circumferential component increases.

matrix. γFe_2O_3 is a modification of Fe_3O_4, magnetite, with Fe and O vacancies. The particles are long needle-like single-domain particles with a length to diameter ratio of about 5 : 1 and a diameter of 0·1 to 1 microns.

In the recording of speech and music it is necessary to have a reasonably linear relation between the change in magnetization and the field applied by the recording head. Clearly, the hysteresis

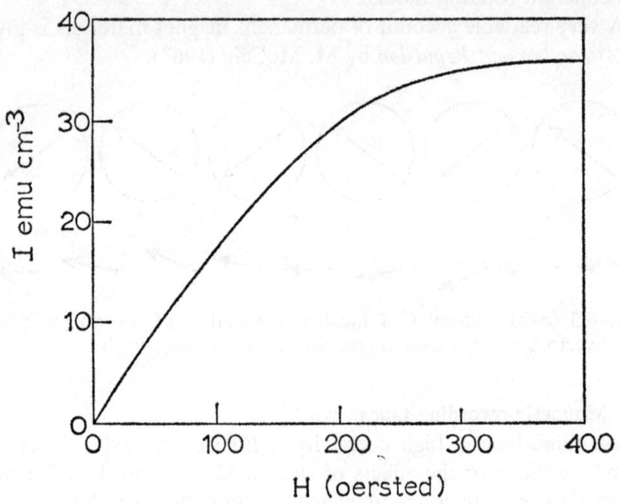

FIG. 6.6 Example of anhysteric magnetization curve; the alternating field is ± 200 oersted.

present in the normal magnetization curve of Fig. 1.2a is to be avoided. Use is made of the fact that a specimen can be brought into a reversible condition by cycling in a field of progressively smaller amplitudes, for example by means of a slowly reducing a.c. field. This is the basis of the standard method of demagnetizing a specimen (Chapter 1). By applying the demagnetizing a.c. field in the presence of a suitable polarizing d.c. field the reversible condition can be obtained anywhere in the magnetization cycle (anhysteretic magnetization); relatively large magnetic fields of the order of the coercive field can then be applied before the change in magnetization becomes irreversible; (the permeability $\mu = dB/dH$ measured under these conditions is the reversible permeability). There is then an approximately linear relation between the polarizing d.c. field and the measured magnetization (Fig. 6.6); the degree of the linearity can be adjusted by control of the field conditions.

In this recording process the signal field, a d.c. field, is applied to the tape simultaneously with the a.c. field which is reduced from a saturation value to zero repeatedly as the signal is recorded. A full description of recording tapes is given in the *Physics of Magnetic Recording* by C. D. Mee.

In the application of recording tapes as memory elements in computers, the conditions for operation are somewhat different in that the information recorded is simply one of positive or negative pulses corresponding to 1 and 0 on the binary system. The conditions for high fidelity are of course not necessary but a high speed of recording is required. Thin films of this nature are used not only on recording tapes but on recording discs as computer stores, but the magnetic characteristics are essentially similar (see § 7.8).

7: Thin Films

Magnetically the important property of a thin film is that it is thin in one direction only and in that direction (z), the demagnetizing factor $N_z = 4\pi$. In the two orthogonal directions x and y in the plane of the film $N_x = N_y = 0$, assuming the film is practically infinite in extent in the xy plane. Thus a field of at least $4\pi I$, approximately 20,000 oersted for iron, is required to magnetize the film to saturation in a direction normal to the film plane. Thus the magnetization lies mainly in the plane of the film except for those films which are highly anisotropic with a preferred axis normal to the film plane.

The study of the domain structure of thin films is somewhat simplified in that to a large extent it is two-dimensional. Experimentally there is the further advantage that the domain configuration observed on the surface can be taken to be that of the interior of the film; if the film is sufficiently thin the domains can be observed by transmission techniques such as Lorentz electron microscopy or Faraday effect so that the internal arrangement of the structure can be observed directly.

Although the mode of formation of the domain structures in thin films follows the same principles as in the bulk material there are significant modifications. Some of these arise, as indicated above, because of the shape factor; others arise because of the mode of preparation, which is usually vacuum deposition and which may introduce a form of anisotropy dependent on the condition of evaporation also on the nature of the substrate. These films are particularly susceptible to the application of a magnetic field present during evaporation; this induces a uniaxial anisotropy with the preferred axis parallel to the field. As the magnetic field need not be large, stray fields are often sufficient. Most evaporated films possess a uniaxial anisotropy arising from this or one of the other effects mentioned above. Variations in orientation of the individual grains also give the film a texture. In addition the structure of the domain boundaries is modified, in the limit for very thin films, with the production of Néel walls instead of Bloch walls. As the presence of the particular type of domain wall may profoundly affect the overall configuration, the problem of the domain wall structure will be considered first.

7.1 Domain walls in thin films – Néel walls

In the analysis of the Bloch wall in section 4.1 no account is taken of the free-pole density arising at the intersection of the wall and the surface. This is justifiable, as in bulk specimens the thickness of the wall δ is much smaller than the thickness of the specimen. In thin films, however, δ and t may well be of the same order of magnitude, and it is necessary to allow for the free-pole effect. Assuming that the domain wall cross-section approximates to be an ellipse with major axes t and δ, the magnetostatic energy (the demagnetizing energy) is then for a Bloch wall (Fig. 7.1a) from eqn (3.4c)

$$E_B = \tfrac{1}{2}N_B I_a{}^2 = \pi\delta I_s{}^2(t + \delta)^{-1} \qquad (7.1)$$

or in terms of energy per unit area of wall $f_B = \pi I_s{}^2\delta^2(t + \delta)^{-1}$. I_a is the average component of the magnetization in the appropriate direction and can be taken as $I_s/\sqrt{2}$ (cf. Fig. 4.1a).

It is obvious from eqn (7.1) that for $t/\delta < 1$ the magnetostatic energy is large. An alternative arrangement of lower energy is one with a magnetic vector in the domain wall rotating in the plane of the film; this is the Néel wall (Fig. 7.1b). The magnetostatic energy per unit volume is then

$$E_N = \tfrac{1}{2}N_N I_a{}^2 = \pi t\, I_s{}^2\, (t + \delta)^{-1} \qquad (7.2)$$

or per unit area of wall $f_N = \pi I_s{}^2\, \delta t\, (t + \delta)^{-1}$.

To derive the wall energy and wall thickness this expression is included in the expression for the wall energy (eqn (4.3)) and the conditions for minimum energy obtained. The wall energy and thickness under the conditions for minimum energy vary with film thickness according to the curves in Fig. 7.2. (Note that the energy of the Néel wall for very small film thickness is identical with that of the Bloch wall for very large values of t, i.e. for bulk specimens. The free-pole energy is very small and the exchange and anisotropy terms are identical in each case.) It will be seen that the critical thickness for the changeover from the Bloch to the Néel wall is below t = δ.

The boundary energy γ in the transition region, because of the large free pole contribution, is high and in practice the expression for γ may be written, for $t/\delta < 1$ and neglecting exchange and anisotropy

$$\gamma = \pi t I_s^2 \tag{7.3}$$

due to the magnetostatic energy (eqn. 7.2).

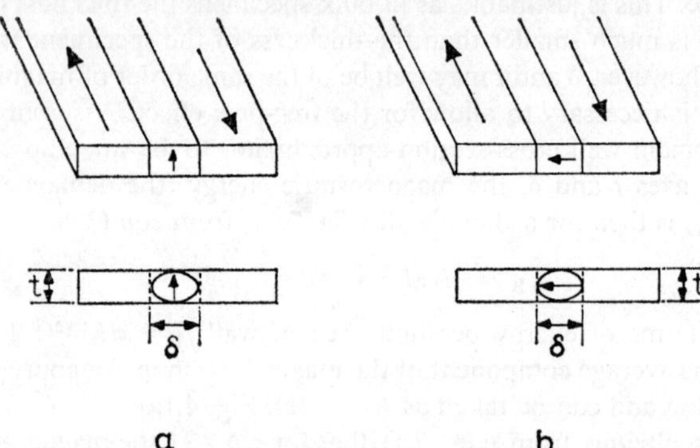

FIG. 7.1 (a–b) Simplified representation of: (a) a Bloch wall and (b) a Néel wall with, below, the wall sections represented by ellipses.

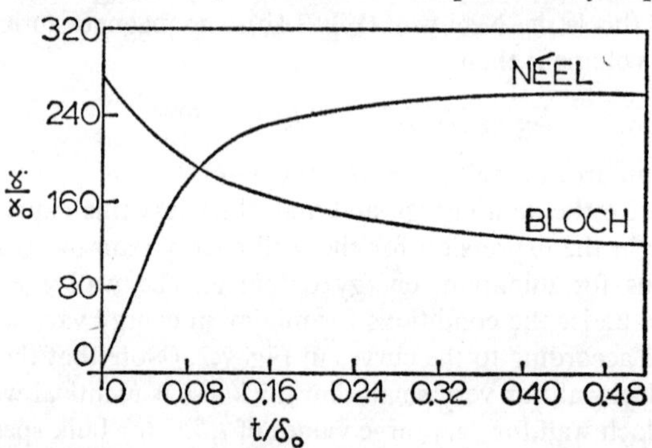

FIG. 7.2 Variation in wall energy for Bloch and Néel walls with specimen thickness t for Permalloy (80 Ni 20 Fe).

In practice the change from the Bloch to the Néel wall is not a sharp transition as it is found that an alternative structure, the cross-tie wall, may be formed in the transition region.

7.2 Cross-tie walls – Bloch lines

The magnetostatic energy associated with a domain wall can be reduced if the direction of the magnetic vectors is reversed periodically. There is thus an alternation in the sign of the free-pole induced on the wall and hence a reduction in the energy. This type of wall formation may have a lower energy than either a Néel or Bloch wall in thin films with a thickness which lies in the transition region discussed above. It can be seen from Fig. 7.3 that when this occurs

FIG. 7.3 Reversal of magnetization vectors in a Néel wall producing a short length of Bloch wall; the diagram also applies to a short length of Néel wall in a Bloch wall.

in a Bloch wall the intervening part of the boundary in which the reversal takes place is, in effect, a short length of a Néel wall. Similarly in a Néel wall there is a short-length Bloch wall which in practice is so short as to be in effect a Bloch line; this is very commonly observed in thin films and is shown in Fig. 7.2.

This configuration is almost invariably modified by the presence of cross-ties by which the effect of stray fields is reduced at alternate Bloch lines (Fig. 7.4, Plate J). It will be seen that the cross-ties form around those Bloch lines where the lines of flux lie antiparallel to the overall direction of magnetization on each side of the main 180° wall.

The energy of a cross-tie wall has been calculated for materials such as Permalloy with low anisotropy and was found to be just over half that of a Néel wall; the critical thickness for the changeover from the Bloch wall to the cross-tie walls is about three times greater than that for the change from the Bloch wall to the Néel wall. This is in agreement with observation that the transition is one

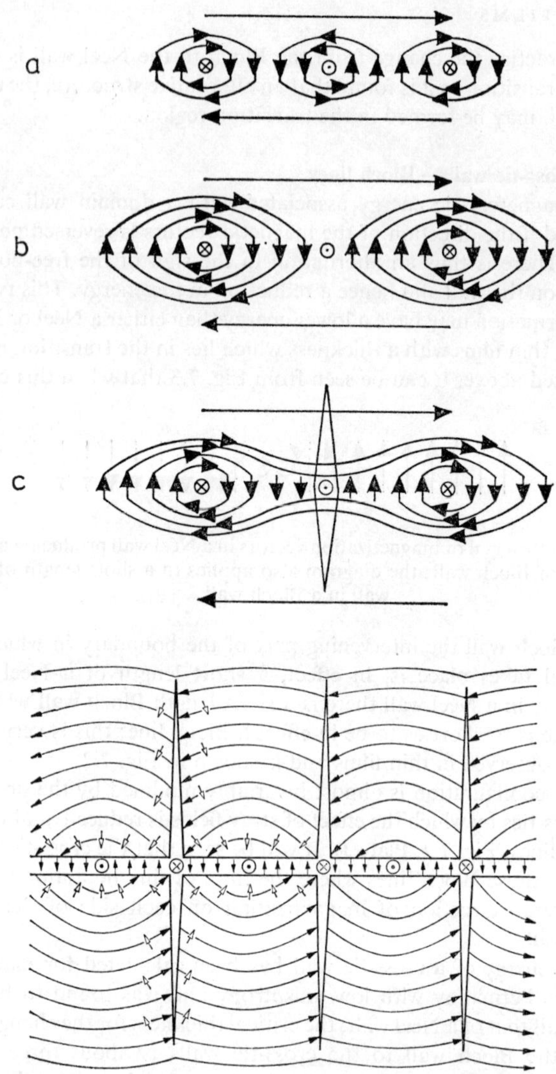

FIG. 7.4 (a–c) Formation of coss-tie wall at a Bloch line in a 180° Néel wall (d) Analysis of cross-tie formation shown in Plate J.

of Bloch/cross-tie/Néel rather than Bloch/Néel direct. In this, as in most domain calculations, there is a discrepancy in that on energetic grounds Néel walls should never be formed whereas they are in fact observed in very thin films.

7.3 The texture of evaporated films – dispersion and ripple

Although most evaporated films possess a uniaxial anisotropy it is found that, from grain to grain, there are local variations in the preferred direction of magnetization; this is known as 'dispersion' of the anisotropy and can be looked upon as a 'magnetic texture'. The effect is particularly noticeable in Lorentz microscopy as it gives rise to a 'ripple' texture in the micrographs with striations lying at right-angles to the direction of magnetization (Plates E, J, K, L, M). This is useful in determining the direction of magnetization in thin films although this can in practice often be deduced from the geometry of the configuration, assuming that the component of the magnetization normal to the boundary is continuous.

An analysis of the effect (based on the approach of Fuller and Hale) is derived from a consideration of Fig. 7.5. The variation in the direction of magnetization giving rise to ripple may be represented in the x and y directions in the plane of the film by

$$I_x = I_s \sin\theta \quad \text{and} \quad I_y = I_s \cos\theta. \tag{7.4a}$$

Consider first the longitudinal ripple

$$\theta = \theta_0 \sin(2\pi y/\lambda) \tag{7.4b}$$

where λ is the wavelength of the ripple and the ripple is a function of y only. The fractional change in the intensity of the image formed at a distance z below the film is, assuming all the angles to be small,

$$\begin{aligned} z\left(\frac{\mathrm{d}\psi_y}{\mathrm{d}y}\right) &= z\frac{\mathrm{d}}{\mathrm{d}y}\left(\frac{4\pi etI_x}{mvc}\right) \\ &= z\frac{\mathrm{d}}{\mathrm{d}y}(\rho I_x) \\ &= z\frac{\mathrm{d}}{\mathrm{d}y}\rho I_s\theta_0 \sin\left(\frac{2\pi y}{\lambda}\right) \\ &= \left(\frac{2\pi z\rho I_s\theta_0}{\lambda}\right)\cos\left(\frac{2\pi y}{\lambda}\right) \end{aligned} \tag{7.5}$$

where $\rho = 4\pi et/mvc$. Similarly for the lateral variation where $\theta = \theta_0 \sin(2\pi x/\lambda)$ the fractional change in intensity is

$$z\left(\frac{d\psi_x}{dx}\right) = z\frac{d}{dx}(\rho I_y)$$

$$= -\left(\frac{\pi z \rho I_s \theta_0^2}{\lambda}\right) \sin\left(\frac{4\pi x}{\lambda}\right). \quad (7.6)$$

Thus the lateral variation is much smaller in amplitude than the longitudinal variation. Equation (7.5) shows that the longitudinal variation gives striations of uniform intensity running at right-angles to the main magnetization I_y.

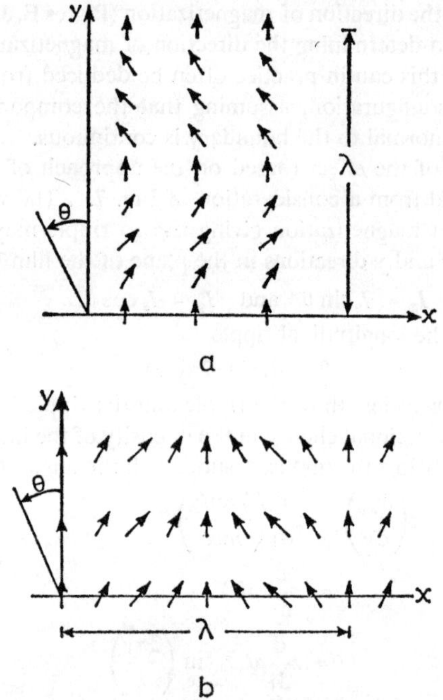

FIG. 7.5 Magnetization ripple in thin films: (a) longitudinal and (b) lateral variation in the direction of magnetization.

7.4 Domain structures in thin films

(i) *Field applied in the easy direction*
In a thin film, unless there is a preferred axis of magnetization normal to the plane of the film, the magnetization lies parallel to the plane of the film and parallel to an easy axis. As the demagnetizing field in the film plane is zero a single domain state is to be expected (eqn 6.5). In practice there are, close to the edge of the film, imperfections and cracks which give rise to localized demagnetizing fields sufficiently large to nucleate (or retain) small subsidiary domains. These domains would be removed or reduced in size on the application of a saturating field but would reappear on reducing the field to zero.

Considering first the effect of a magnetic field in the easy direction of magnetization, the effect of reversing the field after saturation is, of course, to induce the formation of domains with reversed magnetization, i.e. in the same direction as the reverse field. These reversed domains form the enlargement of the reversed edge domains (reverse spikes) which may take place in one of a number of ways (Fig. 7.6).

The edge domains on opposite sides may expand inwards towards each other until they join up to form parallel and antiparallel 180° domains of the usual type. With further increase in field the reverse 180° domains then grow at the expense of the others, with the 180° walls moving in a direction at right angles to the magnetization in the usual way (Fig. 7.6a).

Where edge domains form only on one edge these may grow as above, right across the film to form 180° parallel and antiparallel domains. Alternatively the edge domains may expand towards each other and join up to form a zigzag wall which then moves across the film as a whole (Fig. 7.6b). The zigzag wall is made up of segments which though not parallel to the magnetization lie at only a small angle with it, thus minimizing a free-pole. The wall as a whole, however, lies normal to the magnetization. Zigzag walls are formed particularly where there is some non-uniformity in the field or in the structure of the film (dispersion).

The edge domains may, of course, grow from both sides of the film in this fashion in which case the zigzag walls approach from each side and an island of demagnetization may form; this island will then contract in size as the reverse field is increased (Fig. 7.6c).

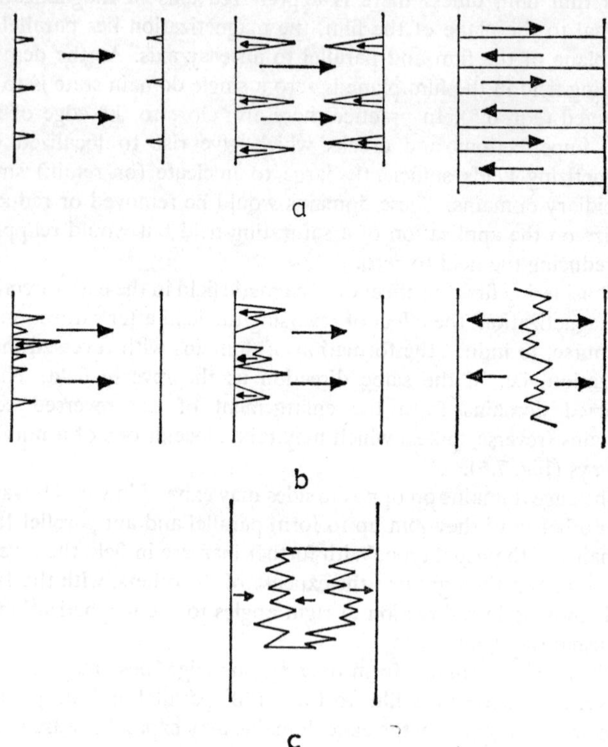

FIG. 7.6 (a–c) Reversal of magnetization in thin film by formation of: (a) reverse 180° domains, (b) zigzag walls, (c) island domain.

(ii) *Field applied in the hard direction*
Where the magnetic field is applied normal to the easy direction but still in the plane of the film, 180° walls are formed at right angles to the field direction (Plate K); at the edge of the specimen 90° closure domains may be formed (as shown in Fig. 5.3c). At higher fields the

change in magnetization takes place by rotation in a similar way to that in the Néel block (Section 5.2).

7.5 Inverted films

So far the domain properties of thin films have been considered in terms in which, on reversing the magnetic field after saturation, the change in magnetization takes place by movement of a 180° domain wall associated with a reversal in the direction of magnetization. When this process is more or less completed further changes in magnetization take place in higher fields by rotation against the anisotropy forces, i.e. the coercivity for wall movement H_c^W is less than H_K. Films can be prepared, however, in which H_c^W is greater than H_K and these are known as inverted films. This effect is associated with a very high dispersion of the magnetization anisotropy so that there is an appreciable variation in the direction of easy magnetization across the film. Plate L shows an inverted thin film which has been subjected to a saturating field so that the magnetization has been constrained to lie in one direction parallel to the field; this is also the overall easy direction (to the right). On reducing the field to zero the magnetization turns towards the local easy direction varying across the film as shown. On reversing the field there is a further rotation of the magnetization and domain walls are formed delineating the regions of more or less uniform magnetization. The magnetization is prevented from rotating further because of the magnetostatic charge produced on the Néel walls which have now been formed. The domain walls are effectively 'locked' and cannot move. It is now necessary to increase the reverse field until either the wholesale reversal of the magnetization within the domains can be brought about or a 180° wall can be nucleated at the edge of the domain. This will then sweep across the whole of the film reversing the magnetization as it moves across the film.

Alternatively, a pair of Bloch lines can be produced in one of the 90° walls and as the field is increased these Bloch lines move apart, breaking up the Néel wall and leaving the region of reversed magnetization. This inversion effect is simply one example of a number of 'anomalous' effects which can be produced in thin films.

7.6 Imperfections in thin films

The interaction between the domain structure and imperfections in thin films follows the general pattern of those observed in bulk specimens with the added effect of the magnetic structure due to dispersion giving rise to ripple and inversion. The effect of inclusions and impurities is to give rise to Néel spikes and similar closure structures even with inclusions as small as 1 micron in diameter

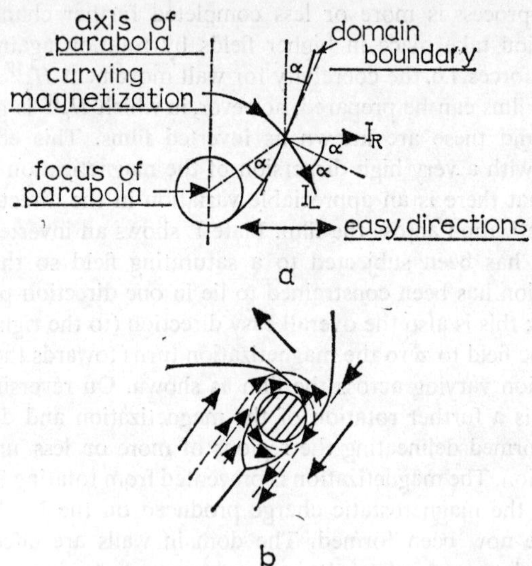

FIG. 7.7 (a–b) (a) Analysis of direction of magnetization at a parabolic domain wall near an inclusion in a thin film, and (b) Drawing of wall configuration (–) actually observed.

(Grundy and Tebble 1968). At very small inclusions, some modification of the domain wall configuration close to the imperfections has been observed. An example of the interaction between the magnetic domain structure at a hole in a thin film is shown in Fig 7.7. This has been analysed by Middelhoek for a film with negligible anisotropy, and the results have been confirmed by experimental observation on thin evaporated films and electropolished metal foils. The domain walls form parabolae with a focus at the centre of the circular hole or non-magnetic inclusion; the parabolic shape arises from the need to have the normal component of the magnetization $\mathbf{n}.\mathbf{I}$ continue across the boundary. In Fig. 7.7 the magnetization on each side of the boundary makes an angle α with the normal to the domain wall so that the normal component is $\cos\alpha$. From the geometry of the figure

$$\frac{dy}{dx} = \cot\alpha = \frac{1}{\tan\alpha} = \frac{1}{t}$$

$$\frac{y}{x} = \cot 2\alpha = \frac{1}{\tan 2\alpha} = \frac{1-t^2}{2t} = \frac{(dy/dx)^2 - 1}{2(dy/dx)}$$

$$2y\left(\frac{dy}{dx}\right) - x\left(\frac{dy}{dx}\right)^2 + x = 0.$$

This is a family of parabolae with vertex at $y = \pm c/2$

$$x^2 = 2cy + c^2 \qquad (7.7)$$

7.7 Single domain particles in thin films

The domain configuration in single domain particles has been studied by Cohen (1965b) in thin films of Permalloy, in the shape of discs about 200 Å thick and a few microns diameter, prepared by a highly involved method of evaporation. The domain structure was examined by Lorentz microscopy, and the configuration predicted for a single domain particle with low anisotropy has been observed and is shown in Plate M with a completely 'circular' magnetization pattern. The two-domain structure does not have the expected closure structure, but the magnetization is rotated round the ends of the walls. The possible structures are shown in Fig. 7.8.

The uniformly magnetized single domain state (Fig. 7.8d) is not observed in the circular particles although the energy is less than that of the two-domain structures of Fig. 7.8b. The energy of the single domain particle associated with the demagnetizing factor N is

$$\tfrac{1}{2}NI_s^2.\pi D^2 t/4 \qquad (7.8)$$

with t the thickness, D the diameter and with $N = \pi^2 t/D$ for $t/D \ll 1$ (cf. eqn 3.4d and section 6.1).

The energy associated with the two-domain particle is made up of the demagnetizing energy (approximately half that of eqn (7.8)) and the wall energy, taken as mainly magnetostatic, $= \pi t I_s^2$ per unit area (eqn (7.3)) giving:

$$\frac{\pi^3 t^2 I_s^2 D}{16} + (\pi t I_s^2) t D = \frac{1{\cdot}3\pi^3 t^2 I_s^2 D}{8} \qquad (7.9)$$

The energy given by eqn (7.9) for the two-domain particle is thus larger than that for the single domain particle (eqn 7.8). This

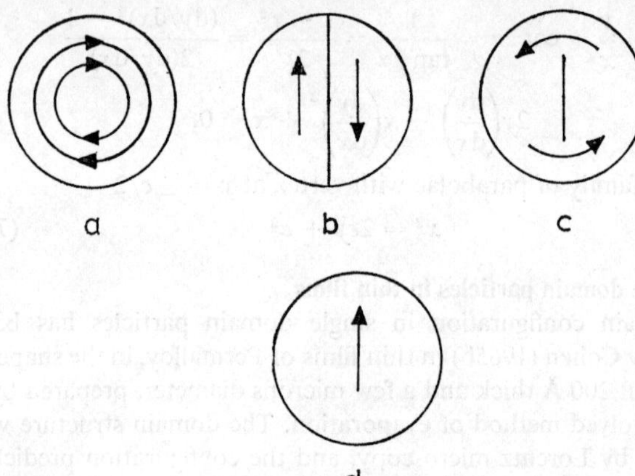

FIG. 7.8 (a–d) Alternative domain configurations in small circular thin films. (cf Plate M).

apparently 'impossible' conclusion is in fact confirmed for circular particles in that a 'modified' two-domain structure is observed.

The formation of the two-domain structure actually observed (Fig. 7.8c) brings about a reduction in the energy of the two-domain state by reducing the wall length and by eliminating the demagnetizing energy. By allowing for the energy associated with the rotation of the magnetization round the end, it is found that the energy of this two-domain structure is reduced by one third and is lower than that of the single domain.

The energy of the circular structure of Fig. 7.8a is due mainly to the magnetocrystalline anisotropy, as the exchange contribution at these dimensions is small, and is given by $\pi KtD^2/8$; this falls off rapidly with D and is thus the preferred structure for small particles.

It is to be noted that the alloy 80% Ni/20% Fe has a low anisotropy and that the completely circular domain structure of Fig. 7.8a might not be observed in a material with a high value of K.

In the specimens with a circular structure a central spot is to be seen, light or dark; this is produced in Lorentz microscopy in the same way as light and dark walls by the convergent or divergent beam; a light or dark spot is formed depending on whether the magnetization curves anti-clockwise or clockwise respectively as viewed from above.

7.8 Thin film applications

There is an increasing interest in the use of thin films as memory elements in computers; these make use of the two stable states represented by the two remanence points $+I_R$ and $-I_R$ on a rectangular hysteresis loop. In the position $+I_R$, for instance, the film magnetization reverses only on the application of a negative field; a positive field has little effect. If we let I_R correspond to 1, then $-I_R$ corresponds to 0 on the binary system so the film can be used as a binary store by the application of switching fields and by picking up the induced e.m.f. on a suitable network of conductors.

The attraction of magnetic thin films lies in the possibility of high switching speeds (less than 1 microsecond) which are possible because eddy-current effects are reduced. Much of the advantage is lost if in the process of the magnetization reversal domain walls are introduced, as the movement of boundaries is a relatively slow process compared with coherent vector rotation.

It will be seen from Chapter 6 that a rectangular loop can be produced by operating a uniaxial film under single domain conditions, but the actual application is not quite as straightforward as might be supposed. The description given below, taken mainly from E. M. Bradley (1960), is of a 'word-organized' or Cambridge-type system consisting of rows of words each word consisting of a column or line of digits. Access to a word or to a single digit must be possible without destroying the information and it must also be possible to record (or change) information by switching individual digits. If the memory consists of a number of magnetic elements arranged in a matrix the problem is how to control the magnetization in each element and to 'sense' the direction of the magnetization without necessarily reversing the magnetization and so destroying the information.

As a preliminary the following elementary properties of a uniaxial single domain thin film will be considered (Fig. 7.9). With the

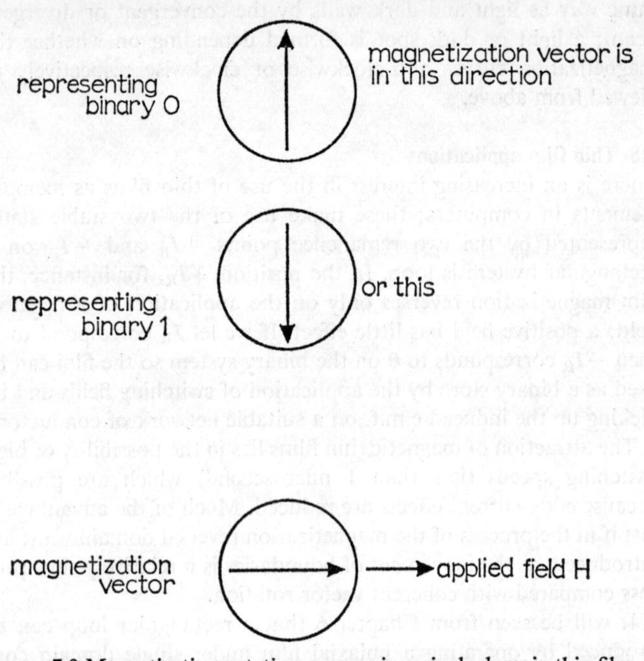

FIG. 7.9 Magnetization rotation process in a single domain thin film.

magnetization parallel to the easy direction, a field is applied in the reverse direction so that a reversal in magnetization takes place, in the ideal case by coherent rotation of the magnetization vector against the anisotropy; on switching off the field the magnetization remains in the new reversed direction. If, however, the field is applied at right angles to the magnetization, i.e. in the hard direction, the magnetization will rotate towards the field until it is nearly aligned with the field. On removing the field the magnetization reverts to its original direction.

Consider a magnetic film at the intersection of two conducting

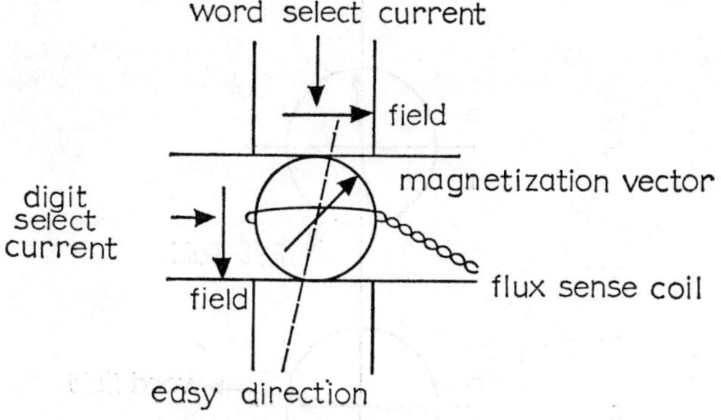

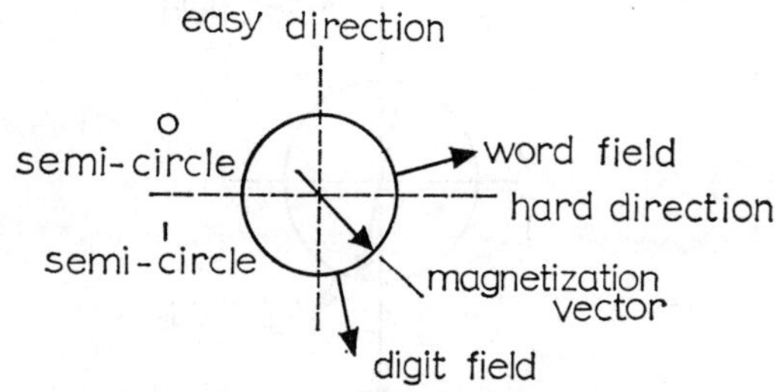

FIG. 7.10 Storage element.

strips as shown in Fig. 7.10. A current through one strip can be used to apply a field at a small angle to the preferred direction; this is the 'digit line' of a 'word-organized' or Cambridge-type system. A current through the other strip causes a field at right angles to that of the digit line, and is thus at a small angle to the 'hard' direction; this is the word-select line. The sense coil is orientated so that it

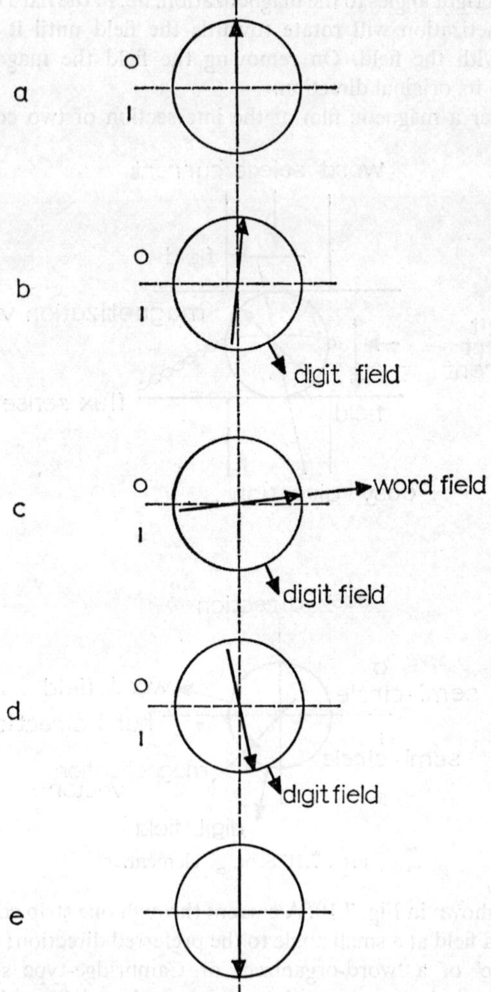

FIG. 7.11 (a–e) Storage element selection process.

'cuts flux' in the direction of the digit field. The arrangement of the fields and their position relative to the preferred direction of the film is shown in Fig. 7.11.

Consider first the 'write' process. At the start of this process the magnetization vector is in the 0 state (Fig. 7.11a). A current is allowed to flow in the 'digit drive' line, if it is required to store a 1 in the chosen element. The field must be less than the coercivity in this direction (Fig. 7.11b) otherwise it could reverse the film by itself. A few millimicroseconds later the 'word' line is activated and the current flowing is sufficient to cause a field which can bring the vector almost into line with it (Fig. 7.11c). The word current is then switched off, leaving the digit current on. The vector can now swing right round almost into line with the digit line field, since the nature of the anisotropy is such that it exerts only a small torque tending to turn the magnetization back to the 0 sense if under 90°, and tending to turn the magnetization to the 1 sense if larger than 90°, i.e. the torque due to the anisotropy changes sign at the 'hard' direction.

The magnetization is now as shown in Fig. 7.11d. The digit field is removed and the magnetization now relaxes into the preferred direction in the 1 sense (Fig. 7.11e). If it had been desired to write a 0, no current would have been passed down the digit line and the magnetization would have been rotated back into the 0 state by the anisotropy torque. The 'read' process is simpler since current is passed only along the word line. If the element was in the 1 state it first rotates nearly into line with the word field and at the end of the pulse relaxes back into the 1 state under the action of the anisotropy torque. If the element was in the 0 state it rotates first in the opposite direction nearly into line with the word-line field, and at the end of the pulse rotates back again to the 0 state. The sense coil is arranged to sense changes of flux in the digit field direction, and different polarity voltage outputs will appear for the two different directions of rotation.

References

DE BLOIS, R. W. (1968) 'Ferromagnetic domains in single crystal Permalloy platelets', *J. Appl. Phys.*, **39**, 442–3.

BRADLEY, E. M. (1960) 'A computer storage matrix using ferromagnetic thin films', *J. Brit. I.R.E.*, 765–84.

BROWN, W. F. (1962) *Magnetostatic Principles in Ferromagnetism.* North Holland, Amsterdam; Wiley–Interscience New York.

COHEN, M. S. (1965) (a) 'Magnetic measurements with Lorentz microscopy *I.E.E.E. Trans. on Magnetics*, Mag. 1, pp. 156-67.

COHEN, M. S. (1965) (b) 'Lorentz microscopy of small ferromagnetic particles', *J. Appl. Phys.*, **36**, 1602–11.

CRAIK, D. J. and TEBBLE, R. S. (1965) *Ferromagnetism and Ferromagnetic Domains.* North Holland, Amsterdam; Wiley–Interscience, New York.

FELDTKELLER, E. (1961) (a) 'Bloch lines in thin nickel iron films', *Symposium of the Electric and Magnetic Properties of Thin Metallic Layers.* Louvain, 98–110.

FELDTKELLER, E. (1961) (b). 'Inverse nickel iron films', *Naturwiss.*, **13**, 474–5.

GRUNDY, P. J. and TEBBLE, R. S. (1968) 'Lorentz electron microscopy', *Advances in Physics*, **17**, 153–242.

SHERWOOD, R. C., REMEIKA, J. P. and WILLIAMS, H. J. (1959). 'Domain behaviour in some magnetic oxides', *J. Appl. Phys.*, **30**, 217–25.

STONER, E. C. and WOHLFARTH, E. P. (1948) 'A mechanism of magnetic hysteresis in heterogeneous alloys', *Phil. Trans. Roy. Soc.*, **A 240**, 599–642.

Literature on magnetic domains

BROWN, W. F. (1962) *Magnetostatic Principles in Ferromagnetism*, North Holland, Amsterdam; Wiley–Interscience, New York.

CAREY, R. and ISAAC, D. J. (1966) *Magnetic Domains and Techniques for Their Observation*, English Universities Press.

CRAIK, D. J. and TEBBLE, R. S. (1961) 'Magnetic domains', *Rep. Prog. Phys.*, **24**, 116–66.

CRAIK, D. J. and TEBBLE, R. S. (1965) *Ferromagnetism and Ferromagnetic Domains*, North Holland, Amsterdam; Wiley–Interscience, New York.

DILLON, J. F. (1963) 'Domains and domain walls'. *In* G. T. RADO and H. SUHL (eds.), *Magnetism*, Academic Press, New York.

GRUNDY, P. J. and TEBBLE, R. S. (1968) 'Lorentz electron miscroscopy', *Advances in Physics*, **17**, 153–242.

KITTEL, C. (1949) 'Physical theory of ferromagnetic domains', ***Rev. Mod. Phys.***, **21**, 541–83.

KITTEL, C. and GALT, J. K. (1965) 'Ferromagnetic domain theory', *Solid State Physics*, **3**, 437–564, Academic Press, New York.

STEWART, K. H. (1954) *Ferromagnetic Domains*. Cambridge University Press.

STONER, E. C. (1950) 'Ferromagnetism—Magnetization curves'. ***Rep. Progr. Physics***, **13**, 83–183.

Other literature

BATES, L. F. (1951) *Modern Magnetism*. Cambridge Unversity Press.

BOZORTH, R. M. (1951) *Ferromagnetism*. Van Nostrand, New York.

CHIKAZUMI, S. (1964) *Physics of Magnetism*. Wiley, New York.

MCCAIG, M. (1967) *Attraction and Repulsion*. Oliver and Boyd, Edinburgh and London.

MEE, C. D. (1964) *The Physics of Magnetic Recording*. Wiley–Interscience, New York.

PRUTTON, M. (1964) *Thin Ferromagnetic Films*. Butterworths, London.

SOOHOO, R. F. (1965) *Magnetic Thin Films*. Harper and Row, New York.

SMIT, J. and WIJN, H. P. J. (1959) *Ferrites*. North Holland, Amsterdam; Wiley, New York.

STANDLEY, G. K. J. (1962) Oxide Magnetic Materials. Clarendon Press, Oxford.

TEBBLE, R. S. and CRAIK, D. J. (1969) *Magnetic Materials*. Wiley–Interscience, New York.

Index